The Maritime Engineering

The Maritime Engineering

Medwin Gale

RANDOM PUBLICATIONS
NEW DELHI (INDIA)

The Maritime Engineering

ISBN 978-93-5111-936-4

Published in 2016 in India by

RANDOM PUBLICATIONS

4376-A/4B, Gali Murari Lal, Ansari Road
New Delhi-110 002
Phone : +9111-43580356, 011-23289044, 011-43142548
e-mail: sales@randompublications.com,
info@randompublications.com, randomexports@gmail.com

Reprinted 2022

Type Setting by : Friends Media, Delhi-110089
Digitally Printed at : Replika Press Pvt. Ltd.

Preface

Marine engineers are skilled individuals involved in the design, construction and maintenance of vehicles and structures used on or around water. Marine engineering technology and techniques are used to create cruise ships, oil platforms and harbors. Marine engineering is the branch of study that deals with the design, development, production and maintenance of the equipments used at sea and on board sea vessels like boats, ships etc. As a matter of fact, it is quite a vast field and it also has many sister arenas like naval architecture and nautical science.

Marine engineering often refers to the engineering of boats, ships, oil rigs and any other marine vessel or structure, but also encompasses oceanographic engineering. Specifically, marine engineering is the discipline of applying engineering sciences, and can include mechanical engineering, electrical engineering, electronic engineering, and computer science, to the development, design, operation and maintenance of watercraft propulsion and also on-board systems and oceanographic technology, not limited to just power and propulsion plants, machinery, piping, automation and control systems etc. for marine vehicles of any kind like surface ships, submarines etc..

Marine propulsion is the mechanism or system used to generate thrust to move a ship or boat across water. While paddles andsails are still used on some smaller boats, most modern ships are propelled by mechanical systems consisting of an electric motor or engine turning a propeller, or less frequently, in pump-jets, an impeller. Marine engineering is the discipline concerned with theengineering design process of marine propulsion systems.

This book is very useful not only for Marine Engineering students but also for research scholars.

– Author

Contents

1

Introduction

Marine engineering often refers to the engineering of boats, ships, oil rigs and any other marine vessel or structure, but also encompasses oceanographic engineering.

Specifically, marine engineering is the discipline of applying engineering sciences, and can include mechanical engineering, electrical engineering, electronic engineering, and computer science, to the development, design, operation and maintenance of watercraft propulsion and also on-board systems and oceanographic technology, not limited to just power and propulsion plants, machinery, piping, automation and control systems etc. for marine vehicles of any kind like surface ships, submarines etc..

The purely mechanical ship operation aspect of marine engineering has some relationship with naval architects. However, whereas naval architects are concerned with the overall design of the ship and its propulsion through the water, marine engineers are focused towards the main propulsion plant, the powering and mechanization aspects of the ship functions such as steering, anchoring, cargo handling, heating, ventilation, air conditioning, electrical power generation and electrical power distribution, interior and exterior communication, and other related requirements.

In some cases, the responsibilities of each industry collide and is not specific to either field. Propellers are examples of one of these types of responsibilities. For naval architects a propeller is a hydrodynamic device. For marine engineers a propeller acts similarly to a pump. Hull vibration, excited by the propeller, is another such area. Noise control and shock hardening must be the joint responsibility of both the naval architect and the marine engineer. In fact, most issues caused by machinery are responsibilities in general.

Not all marine engineering is concerned with moving vessels. Offshore construction, also called offshore engineering, ocean engineering or maritime engineering, is concerned with the technical design of fixed and floating marine structures, such as oil platforms and offshore wind farms.

Fig. Marine Engineers Reviewing Ship Plans.

Oceanographic engineering is concerned with mechanical, electrical, and electronic, and computing technology deployed to support oceanography, and also falls under the umbrella of marine engineering, especially in Britain, where it is covered by the same professional organisation, the IMarEST.

HISTORY OF MARINE ENGINEERING

Archimedes is traditionally regarded as the first marine engineer, having developed a number of marine engineering systems in antiquity. Modern Marine Engineering dates back to the beginning of the Industrial Revolution (early 1700s). In 1712 Thomas Newcomen, a blacksmith, created a steam powered engine to pump water out of mines. In 1807 Robert Fulton successfully used a steam engine to propel a vessel through the water. Fulton's ship used the engine to power a small wooden paddle wheel as its propulsion system. The integration of steam engines into ships was the start of the Marine Engineering profession.

Before the integration of engines in ships, the maritime professions consisted of deck department jobs. This was because there was no engine for engineers to work on. The type of propulsion used before the engine powered paddlewheel was wind power.

Paddle wheel ships were the front runner of the industry for the next 30 years till the next type of propulsion came around. Only twelve years after Fulton's Clermont had her first voyage, the Savannah marked the first sea voyage from America to Europe. Around 50 years later the steam powered paddle wheels had a peak with the creation of the Great Eastern, which was as big as one of the cargo ships of today, 700 feet in length, weighing 22,000 tons. The Great Eastern was said to be ahead of its time and was destined for failure. Since the 1800s there have been many improvements to the design of engines and propellers. The maritime industry holds 90% of all international trade.

Marine Engineers work on more than just engines in ships. Marine engineers are also responsible for building and maintaining offshore oil rigs. These oil rigs were first made by Henry L. Williams in 1896.

CAREER

In 2012, the average annual earnings for Marine Engineers in the U.S. was $96,140 with an average hourly earnings of $46.22.

Education

Maritime universities are dedicated to teaching and training students in maritime professions. Marine Engineers generally have a bachelor's degree in marine engineering, marine engineering technology, or marine systems engineering. Practical training is valued by employers alongside the bachelor's degree.

Unions

There are many different types of unions in the maritime industry. Some unions are specifically for marine engineers, where some are for the maritime industry as a whole.

- Nautilus International
- RMT
- Marine Engineers' Beneficial Association (M.E.B.A) is the longest lasting maritime trade union to date. MEBA was founded in 1875. Its main represented group are licensed maritime officers in both the deck and engine department.
- American Maritime officers (AMO) is the largest union for American maritime officers.

ENGINE ROOM

Fig. Main engine deck of a cargo vessel

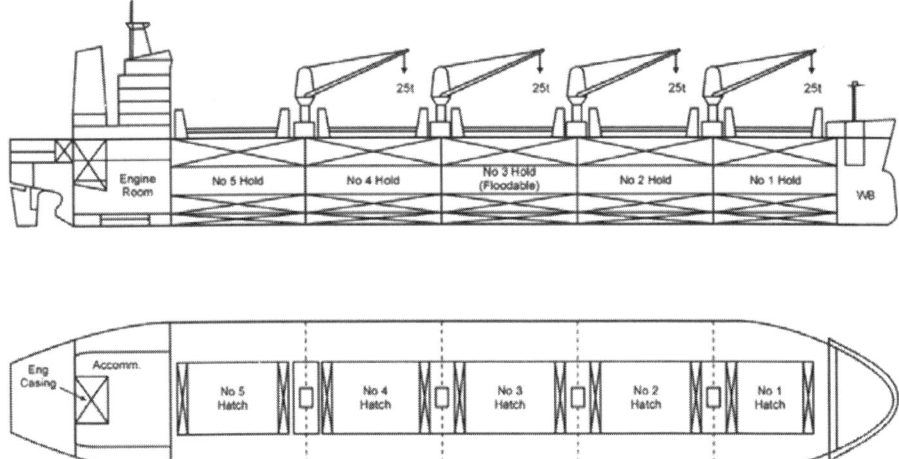

Fig. Location of a ship's engine room on a container ship

On a ship, the engine room, or ER, is the propulsion machinery spaces of the vessel. To increase a vessel's safety and chances of surviving damage, the machinery necessary for operations may be segregated into various spaces. The engine room is one of these spaces, and is generally the largest physical compartment of the machinery space. The engine room houses the vessel's prime mover, usually some variations of a heat engine - diesel engine, gas or steam turbine. On some ships, the machinery space may comprise more than one engine room, such as forward and aft, or port or starboard engine rooms, or may be simply numbered. On a large percentage of vessels, ships and boats, the engine room is located near the bottom, and at the rear, or aft, end of the vessel, and usually comprises few compartments. This design maximizes the cargo carrying capacity of the vessel and situates the prime mover close to the propeller, minimizing equipment cost and problems posed from long shaft lines. The engine room on some ships may be situated mid-ship, especially on vessels built from 1900 to the 1960s. With the increased use of diesel electric propulsion packages, the engine room(s) may be located well forward, low or high on the vessel, depending on the vessel use.

EQUIPMENT

Engines

The engine room of a motor vessel typically contains several engines for different purposes. Main, or propulsion engines are used to turn the ship's propeller and move the ship through the water. They typically burn diesel oil or heavy fuel oil, and may be able to switch between the two. There are many propulsion arrangements for motor vessels, some including multiple engines, propellers, and gearboxes.

Large engines drive electrical generators that provide power for the ship's electrical systems. Large ships typically have three or more synchronized generators to ensure smooth operation. The combined output of a ship's generators is well above the actual power requirement to accommodate maintenance or the loss of one generator.

On a steamship, power for both electricity and propulsion is provided by one or more large boilers giving rise to the alternate name boiler room. High pressure steam from the boiler is used to drive reciprocating engines or turbines for propulsion, and also turbo generators for electricity. Besides propulsion and auxiliary engines, a typical engine room contains many smaller engines, including generators, air compressors, feed pumps, and fuel pumps. Today, these machines are usually powered by small diesel engines or electric motors, but may also use low-pressure steam.

Engine cooling

The engine(s) get required cooling from liquid-to-liquid heat exchangers connected to fresh seawater or divertible to recirculate through tanks of seawater in the engine room. Both supplies draw heat from the engines via the coolant and oil lines. Heat exchangers are plumbed in so that oil is represented by a yellow mark on the flange of the pipes, and relies on paper type gaskets to seal the mating faces of the pipes. Sea water, or brine, is represented by a green mark on the flanges and internal coolant is represented by blue marks on the flanges.

Thrusters

In addition to this array of equipment is the ship's thruster system, typically operated by electric motors controlled from the bridge. These thrusters are laterally mounted propellers that can suck or blow water from port to starboard (i.e. left to right) or vice versa. They are normally used only in maneuvering, e.g. docking operations, and are often banned in tight confines, e.g. drydocks.

Thrusters, like main propellers, are reversible by hydraulic operation. Small embedded hydraulic motors rotate the blades up to 180 degrees to reverse the direction of the thrust.

SAFETY

Fire precautions

Engine rooms are hot, noisy, sometimes dirty, and potentially dangerous. The presence of flammable fuel, high voltage (HV) electrical equipment and internal combustion engines (ICE) means that a serious fire hazard exists in the engine room, which is monitored continuously by the ship's engineering staff and various monitoring systems.

Ventilation

Engine room of the SS *Shieldhall*

If equipped with internal combustion or turbine engines, engine rooms employ some means of providing air for the operation of the engines and associated ventilation. If individuals are normally present in these rooms, additional ventilation should be available to keep engine room temperatures to acceptable limits. If personnel are not normally in the engine space, as in many pleasure boats, the ventilation need only be sufficient to supply the engines with intake air. This would require an unrestricted hull opening of the same size as the intake area of the engine itself, assuming the hull opening is in the engine room itself. Commonly, screens are placed over such openings and if this is done, airflow is reduced by approximately 50%, so the opening area is increased appropriately. The requirement for general ventilation and the requirement for sufficient combustion air are quite different. A typical arrangement might be to make the opening large enough to provide intake air plus 1000 Cubic Feet per Minute (CFM) for additional ventilation. Engines pull sufficient air into the engine room for their own operation. However, additional airflow for ventilation usually requires intake and exhaust blowers.

2

Diesel Engine

We are all familiar with the term diesel and diesel engines but not many of us remember Rudolf Diesel, the brain behind the invention of the diesel engine. Of course I will not go into much detail as this chapter is not meant as a biography but he was a German engineer (where else in the world is engineering such a passion?) who undertook years of experimentation and nearly lost his life when his first engine exploded upon ignition of fuel. The rest as they say ishistory and diesel engines are being successfully used in a variety of applications including diesel marine engines used on different types of ships.

THE DIESEL ENGINE

One of the places where diesel engines play an important role is the shipping industry. Diesel engines are known by the name of compressionignition engines due to technical reasons. There are several ways of classification of diesel engines based on various parameters and some of these classifications are as follows.

- Speed – high speed, medium speed, slow speed
- Usage – automotive engines, locomotive engines, marine engines
- Operation – 2-stroke, 4-stroke, single acting, double acting
- Cylinder arrangement – horizontal, vertical, vee, radial

DIESEL MARINE ENGINES

It can be seen from the above classification that marine engines are those which are used in marine vehicles namelyboats, ships, submarines and so forth. Both 2-stroke as well as 4-stroke engines are used in the marine industry. The engines used for the main propulsion or turning the propeller/s of the normal ships are usually slow speed 2-stroke engines while those used for providing auxiliary power are usually 4-stroke high speed diesel engines.

You can see both these types of engines in the pictures below, the first figure being that of a 2 stroke main propulsionplant while the second is of a 4 stroke generator. Just keep in mind that these images are just for giving you a rough idea otherwise they come in a variety of sizes and flavours.

DIESEL MARINE ENGINE COMPONENTS

The field of diesel marine engines is quite vast and the engine itself is made up of several components.

HISTORY

Fig. A Hornsby-Akroyd oil engineworking at the Great Dorset Steam Fair

In 1885, the English inventor Herbert Akroyd Stuart began investigating the possibility of using paraffin oil (very similar to modern-daydiesel) for an engine, which unlike petrol would be difficult to vaporise in a carburettor as its volatility is not sufficient to allow this.

His hot bulb engines, built from 1891 by Richard Hornsby and Sons, were the first internal combustion engine to use a pressurised fuel injection system. The Hornsby-Akroyd engine used a comparatively low compression ratio, so that the temperature of the air compressed in the combustion chamber at the end of the compression stroke was not high enough to initiate combustion. Combustion instead took place in a separated combustion chamber, the "vaporizer" or "hot bulb" mounted on the cylinder head, into which fuel was sprayed. Self-ignition occurred from contact between the fuel-air mixture and the hot walls of the vaporizer. As the engine's load increased, so did the temperature of the bulb, causing the ignition period to advance; to counteract pre-ignition, water was dripped into the air intake.

The modern Diesel engine incorporates the features of direct (airless) injection and compression-ignition. Both ideas were patented by Akroyd Stuart and Charles Richard Binney in May 1890. Another patent was taken out on 8 October 1890, detailing the working of a complete engine—essentially that of a diesel engine—where air and fuel are introduced separately. The difference between the Akroyd engine and the modern Diesel engine was the requirement to supply extra heat to the cylinder to start the engine from cold. By 1892, Akroyd Stuart had produced an updated version of the engine that no longer required the additional heat source, a year before Diesel's engine.

In 1892, Akroyd Stuart patented a water-jacketed vaporiser to allow compression ratios to be increased. In the same year, Thomas Henry Barton at Hornsbys built a working high-compression version for experimental purposes, whereby the vaporiser was replaced with a cylinder head, therefore not relying on air being preheated, but by combustion through higher compression ratios. It ran for six hours—the first time automatic ignition was produced by compression alone. This was five years before Rudolf Diesel built his well-known high-compression prototype engine in 1897.

Fig. Diesel's original 1897 engine on display at the Deutsches Museum in Munich, Germany

Rudolf Diesel was, however, subsequently credited with the compression ignition engine innovation, despite Akroyd-Stuart's engine being patented two years earlier. The higher compression and thermal efficiency is what distinguishes Diesel's patent, of 3,500 kilopascals (508 psi), from Ackroyd-Stuart's hot bulb compression ignition engine patent, of about 600 kilopascals (87 psi). Diesel improved his engine further, whereas Akroyd Stuart stopped development on his engine in 1893.

In 1892 Diesel received patents in Germany, Switzerland, the United Kingdom and the United States for "Method of and Apparatus for Converting Heat into Work". In 1893 he described a "slow-combustion engine" that first compressed air thereby raising its temperature above the igniting-point of the fuel, then gradually introducing fuel while letting the mixture expand "against resistance sufficiently to prevent an essential increase of temperature and pressure", then cutting off fuel and "expanding without transfer of heat".

In 1894 and 1895 he filed patents and addenda in various countries for his Diesel engine; the first patents were issued in Spain (No. 16,654), France (No. 243,531) and Belgium (No. 113,139) in December 1894, and in Germany (No. 86,633) in 1895 and the United States (No. 608,845) in 1898. He operated his first successful engine in 1897.

At Augsburg, on August 10, 1893, Rudolf Diesel's prime model, a single 10-foot (3.0 m) iron cylinder with a flywheel at its base, ran on its own power for the first time. Diesel spent two more years making improvements and in 1896 demonstrated another model with a theoretical efficiency of 75%, in contrast to the 10% efficiency of the steam engine. By 1898, Diesel had become a millionaire. His engines were used to power pipelines, electric and water plants, automobiles and trucks, and marine craft. They were soon to be used in mines, oil fields, factories, and transoceanic shipping.

TIMELINE

1890s

- *1891:* Herbert Akroyd Stuart invents the first internal combustion engine to use a pressurised fuel injection system.
- *1892:* February 23, Rudolf Diesel obtained a patent (RP 67207) titled "*Arbeitsverfahren und Ausführungsart für Verbrennungsmaschinen*" (Working Methods and Techniques for Internal Combustion Engines).
- *1892:* Akroyd Stuart builds his first working Diesel engine.
- *1893:* Diesel's essay titled *Theory and Construction of a Rational Heat-engine to Replace the Steam Engine and Combustion Engines Known Today* appeared.
- *1893:* August 10, Diesel built his first working prototype in Augsburg.
- 1896 Blackstone & Co, a Stamford farm implement they built lamp start oil engines.
- *1897:* Adolphus Busch licenses rights to the Diesel Engine for the USA and Canada.
- *1898:* Diesel licensed his engine to Branobel, a Russian oil company interested in an engine that could consume non-distilled oil. Branobel's engineers spent four years designing a ship-mounted engine.
- *1899:* Diesel licensed his engine to builders Krupp and Sulzer, who quickly became major manufacturers.

1900s

- *1902:* Until 1910 MAN produced 82 copies of the stationary diesel engine.
- *1903:* Two first diesel-powered ships were launched, both for river and canal operations: *La Petite-Pierre* in France, powered by Dyckhoff-built diesels, and *Vandal* tanker in Russia, powered by Swedish-built diesels with an electrical transmission.
- *1904:* The French built the first diesel submarine, the Z.
- *1905:* Four diesel engine turbochargers and intercoolers were manufactured by Büchl (CH), as well as a scroll-type supercharger from Creux (F) company.

- *1908:* Prosper L'Orange and Deutz developed a precisely controlled injection pump with a needle injection nozzle.
- *1909:* The prechamber with a hemispherical combustion chamber was developed by Prosper L'Orange with Benz.

1910s

- *1910:* The Norwegian research ship *Fram* was a sailing ship fitted with an auxiliary diesel engine, and was thus the first ocean-going ship with a diesel engine.
- *1912:* The Danish built the first ocean-going ship exclusively powered by a diesel engine, MS *Selandia*. The first locomotive with a diesel engine also appeared.
- *1913:* U.S. Navy submarines used NELSECO units. Rudolf Diesel died mysteriously when he crossed the English Channel on the SS *Dresden*.
- *1914:* German U-boats were powered by MAN diesels.
- *1919:* Prosper L'Orange obtained a patent on a prechamber insert and made a needle injection nozzle. First diesel engine from Cummins.

1920s

One of the eight-cylinder 3200 I.H.P. Harland and Wolff—Burmeister & Wain Diesel engines installed in the motorship *Glenapp*. This was the highest powered Diesel engine yet (1920) installed in a ship. Note man standing lower right for size comparison.

- *1921:* Prosper L'Orange built a continuous variable output injection pump.
- *1922:* The first vehicle with a (pre-chamber) diesel engine was Agricultural Tractor Type 6 of the Benz Söhne agricultural tractor OE Benz Sendling.

- *1923:* The first truck with pre-chamber diesel engine made by MAN and Benz. Daimler-Motoren-Gesellschaft testing the first air-injection diesel-engined truck.
- *1924:* The introduction on the truck market of the diesel engine by commercial truck manufacturers in the IAA. Fairbanks-Morse starts building diesel engines.
- *1927:* First truck injection pump and injection nozzles of Bosch. First passenger car prototype of Stoewer.

1930s

- *1930s:* Caterpillar started building diesels for their tractors.
- *1930:* First US diesel-power passenger car (Cummins powered Packard) built in Columbus, Indiana (USA).
- *1930:* Beardmore Tornado diesel engines power the British airship R101.
- *1932:* Introduction of the strongest diesel truck in the world by MAN with 160 hp (120 kW).
- *1933:* First European passenger cars with diesel engines (Citroën Rosalie); Citroën used an engine of the English diesel pioneer SirHarry Ricardo. The car did not go into production due to legal restrictions on the use of diesel engines.
- *1934:* First turbo diesel engine for a railway train by Maybach. First streamlined, stainless steel passenger train in the US, the Pioneer Zephyr, using a Winton engine.
- *1934:* First tank equipped with diesel engine, the Polish 7TP.
- *1934–35:* Junkers Motorenwerke in Germany started production of the Jumo aviation diesel engine family, the most famous of these being the Jumo 205, of which over 900 examples were produced by the outbreak of World War II.
- *1936:* Mercedes-Benz built the 260D diesel car. AT&SF inaugurated the diesel train Super Chief. The airship Hindenburg was powered by diesel engines. First series of passenger cars manufactured with diesel engine (Mercedes-Benz 260 D, Hanomag andSaurer). Daimler Benz airship diesel engine 602LOF6 for the LZ129 *Hindenburg* airship.
- *1937:* The Soviet Union developed the Kharkiv model V-2 diesel engine, later used in the T-34 tanks, widely regarded as the best tank chassis of World War II.
- *1937:* BMW 114 experimental airplane diesel engine development.
- *1938:* General Motors forms the GM Diesel Division, later to become Detroit Diesel, and introduces the Series 71 inline high-speed medium-horsepower two stroke engine; GM's EMD subsidiary

introduces the 567 two stroke medium-speed high-horsepower engine for locomotive, ship and stationary applications; These GM and EMD engines utilize GM's patented Unit injector.

- *1938:* First turbo diesel engine of Saurer.

1940s

- *1942:* Tatra started production of Tatra 111 with air-cooled V12 diesel engine.
- *1943–46:* The common-rail (CRD) system was invented (and patented by) Clessie Cummins
- *1944:* Development of air cooling for diesel engines by Klöckner Humboldt Deutz AG (KHD) for the production stage, and later also for Magirus Deutz.

1950s

- *1953:* Turbo-diesel truck for Mercedes in small series.
- *1954:* Turbo-diesel truck in mass production by Volvo. First diesel engine with an overhead cam shaft of Daimler Benz.
- 1958 EMD introduces turbocharging for its 567 series of medium speed, high horsepower locomotive, stationary and marine engines. Every subsequent engine (645 and 710) would incorporate this turbocharger.

1960s

- *1960:* The diesel drive displaced steam turbines and coal fired steam engines.
- *1962–65:* A diesel compression braking system, eventually to be manufactured by Jacobs (of drill chuck fame) and nicknamed the "Jake Brake", was invented and patented by Clessie Cummins.
- *1968:* Peugeot introduced the first 204 small cars with a transversally mounted diesel engine and front-wheel drive.

1970s

- *1973:* DAF produced an air-cooled diesel engine.
- *1976 February:* Tested a diesel engine for the Volkswagen Golf passenger car. The Cummins Common Rail injection system was further developed by the ETH Zurich from 1976 to 1992.
- *1978:* Mercedes-Benz produced the first passenger car with a turbo-diesel engine (Mercedes-Benz 300 SD). Oldsmobile introduced the first passenger car diesel engine produced by an American car company.
- *1979:* Peugeot 604, the first turbo-diesel car to be sold in Europe.

1980s

- *1985:* ATI Intercooler diesel engine from DAF. European Truck Common Rail system with the IFA truck type W50 introduced.
- *1986:* BMW 524td, the world's first passenger car equipped with an electronically controlled injection pump (developed by Bosch). The same year, the Fiat Croma was the first passenger car in the world to have a direct injection (turbocharged) diesel engine.
- *1987:* Most powerful production truck with a 460 hp (340 kW) MAN diesel engine.
- *1989:* Audi 100, the first passenger car in the world with a turbocharged direct injection and electronic control diesel engine.

1990s

- *1991:* European emission standards Euro 1 met with the truck diesel engine of Scania.
- *1993:* Pump nozzle injection introduced in Volvo truck engines.
- *1994:* Unit injector system by Bosch for diesel engines. Mercedes-Benz unveils the first automotive diesel engine with four valves per cylinder. Medium speed high horsepower locomotive, ship and stationary diesel engines have utilized four valves per cylinder since at least 1938.
- *1995:* First successful use of common rail in a production vehicle, by Denso in Japan, Hino "Rising Ranger" truck.
- *1996:* First diesel engine with direct injection and four valves per cylinder, used in the Opel Vectra.
- *1997:* First common rail diesel engine in a passenger car, the Alfa Romeo 156.
- *1998:* BMW made history by winning the 24 Hour Nürburgring race with the 320d, powered by a two-litre, four-cylinder diesel engine. The combination of high-performance with better fuel efficiency allowed the team to make fewer pit stops during the long endurance race. Volkswagen introduces three and four-cylinder turbodiesel engines, with Bosch-developed electronically controlled unit injectors. Smart presented the first common rail three-cylinder diesel engine used in a passenger car (the Smart City Coupé).
- *1999:* Euro 3 of Scania and the first common rail truck diesel engine of Renault.

2000s

- *2002:* A street-driven Dodge Dakota pickup with a 735 horsepower (548 kW) diesel engine built at Gale banks engineering hauls its own service trailer to the Bonneville Salt Flats and set an FIA land speed record as the world's fastest pickup truck with a one-way run of 222 mph (357 km/h) and a two-way average of 217 mph (349 km/h).

- *2004:* In Western Europe, the proportion of passenger cars with diesel engine exceeded 50%. Selective catalytic reduction (SCR) system in Mercedes, Euro 4 with EGR system and particle filters of MAN. Piezoelectric injector technology by Bosch.
- *2006:* Audi R10 TDI won the 12 Hours of Sebring and defeated all other engine concepts. The same car won the 2006 24 Hours of Le Mans. Euro 5 for all Iveco trucks. JCB Dieselmax broke the FIA diesel land speed record from 1973, eventually setting the new record at over 350 mph (563 km/h).
- *2007:* Lombardini develops a new 440 cc twin-cyinder common rail diesel engine, which two years later sees application in automotive use, in the Ligier microcars. At the time, this engine was considered to be the smallest twin-cyinder engine with a common rail system.
- *2008:* Subaru introduced the first horizontally opposed diesel engine to be fitted to a passenger car. This is a Euro 5 compliant engine with an EGR system. SEAT wins the drivers' title and the manufacturers' title in the FIA World Touring Car Championship with the SEAT León TDI. The achievements are repeated in the following season.
- *2009:* Volkswagen won the 2009 Dakar Rally held in Argentina and Chile. The first diesel to do so. Race Touareg 2 models finished first and second. The same year, Volvo is claimed the world's strongest truck with their FH16 700. An inline 6-cylinder, 16 L (976 cu in) 700 hp (522 kW) diesel engine producing 3150 Nm (2323.32 lb•ft) of torque and fully complying with Euro 5 emission standards.

2010s

- *2010:* Mitsubishi developed and started mass production of its 4N13 1.8 L DOHC I4, the world's first passenger car diesel engine that features a variable valve timingsystem. Scania AB's V8 had the highest torque and power ratings of any truck engine, 730 hp (544 kW) and 3,500 N·m (2,581 ft·lb).
- *2011:* Piaggio launches a twin-cyinder turbodiesel engine, with common rail injection, on its new range of microvans.

OPERATING PRINCIPLE

p-V Diagram for the Ideal Diesel cycle. The cycle follows the numbers 1–4 in clockwise direction. The horizontal axis is Volume of the cylinder.

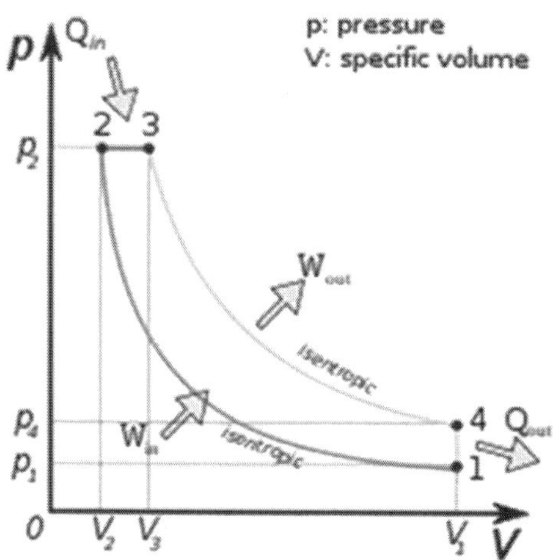

In the diesel cycle the combustion occurs at almost constant pressure. On this diagram the work that is generated for each cycle corresponds to the area within the loop.

Diesel engine model, left side

Diesel engine model, right side

The diesel internal combustion engine differs from the gasoline powered Otto cycle by using highly compressed hot air to ignite the fuel rather than using a spark plug (*compression ignition* rather than *spark ignition*).

In the true diesel engine, only air is initially introduced into the combustion chamber. The air is then compressed with a compression ratio typically between 15:1 and 23:1 resulting in 40-bar (4.0 MPa; 580 psi) pressure compared to 8 to 14 bars (0.80 to 1.40 MPa; 120 to 200 psi) in the petrol engine. This high compression causes the temperature of the air to rise to 550 °C (1,022 °F). At about the top of the compression stroke, fuel is injected directly into the compressed air in the combustion chamber. This may be into a (typically toroidal) void in the top of the piston or a *pre-chamber* depending upon the design of the engine. The fuel injector ensures that the fuel is broken down into small droplets, and that the fuel is distributed evenly. The heat of the compressed air vaporizes fuel from the surface of the droplets. The vapour is then ignited by the heat from the compressed air in the combustion chamber, the droplets continue to vaporise from their surfaces and burn, getting smaller, until all the fuel in the droplets has been burnt. Combustion occurs at a substantially constant pressure during the initial part of the power stroke. The start of vaporisation causes a delay before ignition and the characteristic diesel knocking sound as the vapour reaches ignition temperature and causes an abrupt increase in pressure above the piston (not shown on the P-V indicator diagram). When combustion is complete the combustion gases expand as the piston descends further; the high pressure in the cylinder drives the piston downward, supplying power to the crankshaft.

As well as the high level of compression allowing combustion to take place without a separate ignition system, a high compression ratiogreatly increases the engine's efficiency. Increasing the compression ratio in a spark-ignition engine where fuel and air are mixed before entry to the cylinder is limited by the need to prevent damaging pre-ignition. Since only air is compressed in a diesel engine, and fuel is not introduced into the cylinder until shortly before top dead centre (TDC), premature detonation is not a problem and compression ratios are much higher.

The p–V diagam is a simplified and idealised representation of the events involved in a Diesel engine cycle, arranged to illustrate the similarity with a Carnot cycle. Starting at 1, the piston is at bottom dead centre and both valves are closed at the start of the compression stroke; the cylinder contains air at atmospheric pressure. Between 1 and 2 the air is compressed adiabatically—that is without heat transfer to or from the environment—by the rising piston. (This is only approximately true since there will be some heat exchange with the cylinder walls.) During this compression, the volume is reduced, the pressure and temperature both rise. At or slightly before 2 (TDC) fuel is injected and burns in the compressed hot air. Chemical energy is released and this constitutes an injection of thermal energy (heat) into the compressed gas. Combustion and heating occur between 2 and 3. In this interval the pressure remains constant since the piston descends, and the volume increases; the temperature rises as a consequence of the energy of combustion. At 3 fuel injection and combustion are complete, and the cylinder contains gas at a higher temperature than at 2. Between 3 and 4 this hot gas expands, again approximately adiabatically. Work is done on the system to which the engine is connected. During this expansion phase the volume of the gas rises, and its temperature and pressure both fall. At 4 the exhaust valve opens, and the pressure falls abruptly to atmospheric (approximately). This is unresisted expansion and no useful work is done by it. Ideally the adiabatic expansion should continue, extending the line 3–4 to the right until the pressure falls to that of the surrounding air, but the loss of efficiency caused by this unresisted expansion is justified by the practical difficulties involved in recovering it (the engine would have to be much larger). After the opening of the exhaust valve, the exhaust stroke follows, but this (and the following induction stroke) are not shown on the diagram. If shown, they would be represented by a low-pressure loop at the bottom of the diagram. At 1 it is assumed that the exhaust and induction strokes have been completed, and the cylinder is again filled with air. The piston-cylinder system absorbs energy between 1 and 2—this is the work needed to compress the air in the cylinder, and is provided by mechanical kinetic energy stored in the flywheel of the engine. Work output is done by the piston-cylinder combination between 2 and 4. The difference between these two increments of work is the indicated work output per cycle, and is represented by the area enclosed by the p–V loop. The adiabatic expansion is in a higher pressure range than that of the compression because the gas in the cylinder is hotter during expansion than during compression. It is for this reason that the loop has a finite area, and the net output of work during a cycle is positive.

Early fuel injection systems

Diesel's original engine injected fuel with the assistance of compressed air, which atomized the fuel and forced it into the engine through a nozzle (a similar principle to an aerosol spray). The nozzle opening was closed by a pin valve lifted by the camshaft to initiate the fuel injection before top dead centre (TDC). This is called an air-blast injection. Driving the three stage compressor used some power but the efficiency and net power output was more than any other combustion engine at that time.

Diesel engines in service today raise the fuel to extreme pressures by mechanical pumps and deliver it to the combustion chamber by pressure-activated injectors without compressed air. With direct injected diesels, injectors spray fuel through 4 to 12 small orifices in its nozzle. The early air injection diesels always had a superior combustion without the sharp increase in pressure during combustion. Research is now being performed and patents are being taken out to again use some form of air injection to reduce the nitrogen oxides and pollution, reverting to Diesel's original implementation with its superior combustion and possibly quieter operation. In all major aspects, the modern diesel engine holds true to Rudolf Diesel's original design, that of igniting fuel by compression at an extremely high pressure within the cylinder. With much higher pressures and high technology injectors, present-day diesel engines use the so-called solid injection system applied by Herbert Akroyd Stuart for his hot bulb engine. The indirect injection engine could be considered the latest development of these low speed *hot bulb* ignition engines.

Fuel delivery

Diesel engines are also produced with two significantly different injection locations. "Direct" and "Indirect". Indirect injected engines place the injector in a pre-combustion chamber in the head which due to thermal losses generally require a "glow plug" to start and very high compression ratio. Usually in the range of 21:1 to 23:1 ratio. Direct injected engines use a generally donut shaped combustion chamber void on the top of the piston. Thermal efficiency losses are significantly lower in DI engines which facilitates a much lower compression ratio generally between 14:1 and 20:1 but most DI engines are closer to 17:1. The direct injected process is significantly more internally violent and thus requires careful design, and more robust construction. The lower compression ratio also creates challenges for emissions due to partial burn. Turbocharging is particularly suited to DI engines since the low compression ratio facilitates meaningful forced induction, and the increase in airflow allows capturing additional fuel efficiency not only from more complete combustion, but also from lowering parasitic efficiency losses when properly operated, by widening both power and efficiency curves. The violent combustion process of direct injection also creates more noise, but modern designs using "split shot" injectors or similar multi shot processes have dramatically amended this issue by firing a small charge of fuel before the main delivery which pre-charges the combustion chamber for a less abrupt and in most cases slightly cleaner burn.

A vital component of all diesel engines is a mechanical or electronic governor which regulates the idling speed and maximum speed of the engine by controlling the rate of fuel delivery. Unlike Otto-cycle engines, incoming air is not throttled and a diesel engine without a governor cannot have a stable idling speed and can easily overspeed, resulting in its destruction. Mechanically governed fuel injection systems are driven by the engine's gear train. These systems use a combination of springs and weights to control fuel delivery relative to both load and speed. Modern electronically controlled diesel engines control fuel delivery by use of an electronic control module (ECM) or electronic control unit (ECU). The ECM/ECU receives an engine speed signal, as well as other operating parameters such as intake manifold pressure and fuel temperature, from a sensor and controls the amount of fuel and start of injection timing through actuators to maximise power and efficiency and minimise emissions. Controlling the timing of the start of injection of fuel into the cylinder is a key to minimizing emissions, and maximizing fuel economy (efficiency), of the engine. The timing is measured in degrees of crank angle of the pistonbefore top dead centre. For example, if the ECM/ECU initiates fuel injection when the piston is 10° before TDC, the start of injection, or timing, is said to be 10° BTDC. Optimal timing will depend on the engine design as well as its speed and load, and is usually 4° BTDC in 1,350–6,000 HP, net, "medium speed" locomotive, marine and stationary diesel engines.

Advancing the start of injection (injecting before the piston reaches to its SOI-TDC) results in higher in-cylinder pressure and temperature, and higher efficiency, but also results in increased engine noise due to faster cylinder pressure rise and increased oxides of nitrogen (NO_x) formation due to higher combustion temperatures. Delaying start of injection causes incomplete combustion, reduced fuel efficiency and an increase in exhaust smoke, containing a considerable amount of particulate matter and unburnedhydrocarbons.

Major advantages

Diesel engines have several advantages over other internal combustion engines:

- They burn less fuel than a petrol engine performing the same work, due to the engine's higher temperature of combustion and greater expansion ratio. Gasoline engines are typically 30% efficient while diesel engines can convert over 45% of the fuel energy into mechanical energy.
- They have no high voltage electrical ignition system, resulting in high reliability and easy adaptation to damp environments. The absence of coils, spark plug wires, etc., also eliminates a source of radio frequency emissions which can interfere with navigation and communication equipment, which is especially important in marine and aircraft applications, and for preventing interference with radio telescopes.
- The longevity of a diesel engine is generally about twice that of a petrol engine due to the increased strength of parts used. Diesel fuel has better lubrication properties than petrol as well. Indeed, in unit injectors, the fuel is employed for three distinct purposes: injector lubrication, injector cooling and injection for combustion.
- Diesel fuel is distilled directly from petroleum. Distillation yields some gasoline, but the yield would be inadequate without catalytic reforming, which is a more costly process.
- Diesel fuel is considered safer than petrol in many applications.

Although diesel fuel will burn in open air using a wick, it will not explode and does not release a large amount of flammable vapor. The low vapor pressure of diesel is especially advantageous in marine applications, where the accumulation of explosive fuel-air mixtures is a particular hazard. For the same reason, diesel engines are immune to vapor lock.

- For any given partial load the fuel efficiency (mass burned per energy produced) of a diesel engine remains nearly constant, as opposed to petrol and turbine engines which use proportionally more fuel with partial power outputs.
- They generate less waste heat in cooling and exhaust.
- Diesel engines can accept super- or turbo-charging pressure without any natural limit, constrained only by the strength of engine components. This is unlike petrol engines, which inevitably suffer detonation at higher pressure.
- The carbon monoxide content of the exhaust is minimal.
- Biodiesel is an easily synthesized, non-petroleum-based fuel (through transesterification) which can run directly in many diesel engines, while gasoline engines either need adaptation to run synthetic fuels or else use them as an additive to gasoline (e.g., ethanol added to gasohol).

Mechanical and electronic injection

Many configurations of fuel injection have been used over the course of the 20th century.

Most present-day diesel engines use a mechanical single plunger high-pressure fuel pump driven by the engine crankshaft. For each engine cylinder, the corresponding plunger in the fuel pump measures out the correct amount of fuel and determines the timing of each injection. These engines use injectors that are very precise spring-loaded valves that open and close at a specific fuel pressure. Separate high-pressure fuel lines connect the fuel pump with each cylinder. Fuel volume for each single combustion is controlled by a slanted groove in the plunger which rotates only a few degrees releasing the pressure and is controlled by a mechanical governor, consisting of weights rotating at engine speed constrained by springs and a lever. The injectors are held open by the fuel pressure. On high-speed engines the plunger pumps are together in one unit. The length of fuel lines from the pump to each injector is normally the same for each cylinder in order to obtain the same pressure delay.

A cheaper configuration on high-speed engines with fewer than six cylinders is to use an axial-piston distributor pump, consisting of one rotating pump plunger delivering fuel to a valve and line for each cylinder (functionally analogous to points and distributor cap on an Otto engine).

Many modern systems have a single fuel pump which supplies fuel constantly at high pressure with a common rail (single fuel line common) to each injector. Each injector has asolenoid operated by an electronic control unit, resulting in more accurate control of injector opening times that depend on other control conditions, such as engine speed and loading, and providing better engine performance and fuel economy.

Both mechanical and electronic injection systems can be used in either direct or indirect injection configurations.

Two-stroke diesel engines with mechanical injection pumps can be inadvertently run in reverse, albeit in a very inefficient manner, possibly damaging the engine.Large ship two-stroke diesels are designed to run in either direction, obviating the need for a gearbox.

Indirect injection

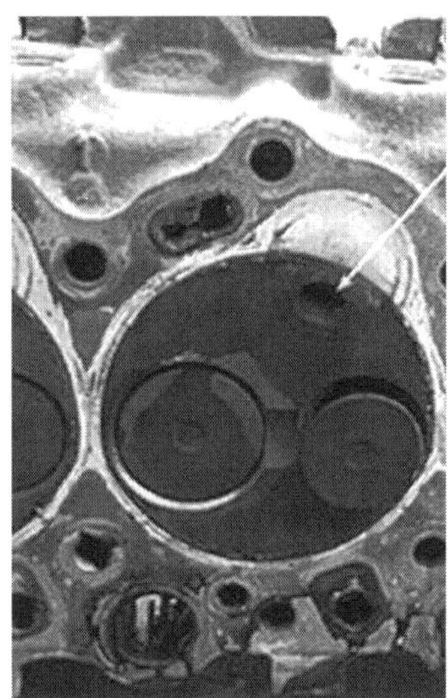

Arrow indicates opening from pre-chamber

An indirect injection diesel engine delivers fuel into a chamber off the combustion chamber, called a pre combustion chamber or ante-chamber, where combustion begins and then spreads into the main combustion chamber, assisted by turbulence created in the chamber. This system allows for a smoother, quieter running engine, and because combustion is assisted by turbulence, injector pressures can be lower, about 100 bar (10 MPa; 1,500 psi), using a single orifice tapered jet injector. Mechanical injection systems allowed high-speed running suitable for road vehicles (typically up to speeds of around 4,000 rpm).

The pre-chamber had the disadvantage of increasing heat loss to the engine's cooling system, and restricting the combustion burn, which reduced the efficiency by 5–10%. Indirect injection engines are cheaper to build and it is easier to produce smooth, quiet-running vehicles with a simple mechanical system. In road-going vehicles most prefer the greater efficiency and better controlled emission levels of direct injection. Indirect injection diesels can still be found in the many ATV diesel applications.

Direct injection

Direct injection diesel engines have injectors mounted at the top of the combustion chamber. The injectors are activated using one of two methods - hydraulic pressure from the fuel pump, or an electronic signal from an engine controller.

Hydraulic pressure activated injectors can produce harsh engine noise. Fuel consumption is about 15–20% lower than indirect injection diesels. The extra noise is generally not a problem for industrial uses of the engine, but for automotive usage, buyers have to decide whether or not the increased fuel efficiency would compensate for the extra noise. The noise level of modern common rail direct injection diesel engines with multiple injections per combustion cycle is however comparable to those of their gasoline counterparts.

Electronic control of the fuel injection transformed the direct injection engine by allowing much greater control over the combustion.

Unit direct injection

Unit direct injection also injects fuel directly into the cylinder of the engine. In this system the injector and the pump are combined into one unit positioned over each cylinder controlled by the camshaft. Each cylinder has its own unit eliminating the high-pressure fuel lines, achieving a more consistent injection. This type of injection system, also developed by Bosch, is used by Volkswagen AG in cars (where it is called a *Pumpe-Düse-System*—literally *pump-nozzle system*) and by Mercedes-Benz ("PLD") and most major diesel engine manufacturers in large commercial engines (MAN SE, CAT, Cummins, Detroit Diesel, Electro-Motive Diesel, Volvo).

With recent advancements, the pump pressure has been raised to 2,400 bars (240 MPa; 35,000 psi), allowing injection parameters similar to common rail systems.

Common rail direct injection

In common rail systems, the separate pulsing high-pressure fuel line to each cylinder's injector is also eliminated. Instead, a high-pressure pump pressurizes fuel at up to 2,500 bar (250 MPa; 36,000 psi), in a "common rail". The common rail is a tube that supplies each computer-controlled injector containing a precision-machined nozzle and a plunger driven by a solenoid or piezoelectric actuator.

COLD WEATHER

Starting

In cold weather, high speed diesel engines can be difficult to start because the mass of the cylinder block and cylinder head absorb the heat of compression, preventing ignition due to the higher surface-to-volume ratio. Pre-chambered engines make use of small electric heaters inside the pre-chambers called glowplugs, while direct-injected engines have these glowplugs in the combustion chamber.

Many engines use resistive heaters in the intake manifold to warm the inlet air for starting, or until the engine reaches operating temperature. Engine block heaters (electric resistive heaters in the engine block) connected to the utility grid are used in cold climates when an engine is turned off for extended periods (more than an hour), to reduce startup time and engine wear. Block heaters are also used for emergency power standby Diesel-powered generators which must rapidly pick up load on a power failure. In the past, a wider variety of cold-start methods were used. Some engines, such as Detroit Diesel engines used a system to introduce small amounts of ether into the inlet manifold to start combustion. Others used a mixed system, with a resistive heater burning methanol. An impromptu method, particularly on out-of-tune engines, is to manually spray an aerosol can of ether-based engine starter fluid into the intake air stream (usually through the intake air filter assembly).

Gelling

Diesel fuel is also prone to *waxing* or *gelling* in cold weather; both are terms for the solidification of diesel oil into a partially crystalline state. The crystals build up in the fuel line (especially in fuel filters), eventually starving the engine of fuel and causing it to stop running. Low-output electric heaters in fuel tanks and around fuel lines are used to solve this problem. Also, most engines have a *spill return* system, by which any excess fuel from the injector pump and injectors is returned to the fuel tank. Once the engine has warmed, returning warm fuel prevents waxing in the tank.

Due to improvements in fuel technology with additives, waxing rarely occurs in all but the coldest weather when a mix of diesel and kerosene may be used to run a vehicle. Gas stations in regions with a cold climate are required to offer winterized diesel in the cold seasons that allow operation below a specific Cold Filter Plugging Point. In Europe these diesel characteristics are described in the EN 590 standard.

SUPERCHARGING AND TURBOCHARGING

Most diesels are now turbocharged and some are both turbo charged and supercharged. Because diesels do not have fuel in the cylinder before combustion is initiated, more than one bar (100 kPa) of air can be loaded in the cylinder without preignition. A turbocharged engine can produce significantly more power than a naturally aspirated engine of the same configuration, as having more air in the cylinders allows more fuel to be burned and thus more power to be produced. A supercharger is powered mechanically by the engine's crankshaft, while a turbocharger is powered by the engine exhaust, not requiring any mechanical power. Turbocharging can improve the fuel economy of diesel engines by recovering waste heat from the exhaust, increasing the excess air factor, and increasing the ratio of engine output to friction losses.

A two-stroke engine does not have a discrete exhaust and intake stroke and thus is incapable of self-aspiration. Therefore, all two-stroke engines must be fitted with a blower to charge the cylinders with air and assist in dispersing exhaust gases, a process referred to as scavenging. In some cases, the engine may also be fitted with a turbocharger, whose output is directed into the blower inlet. A few designs employ a hybrid turbocharger (a turbo-compressor system) for scavenging and charging the cylinders, which device is mechanically driven at cranking and low speeds to act as a blower, but which acts as a true turbocharger at higher speeds and loads. A hybrid turbocharger can revert to compressor mode during commands for large increases in engine output power.

As turbocharged or supercharged engines produce more power for a given engine size as compared to naturally aspirated engines, attention must be paid to the mechanical design of components, lubrication, and cooling to handle the power. Pistons are usually cooled with lubrication oil sprayed on the bottom of the piston. Large engines may use water, sea water, or oil supplied through telescoping pipes attached to the crosshead.

TYPES

Size groups

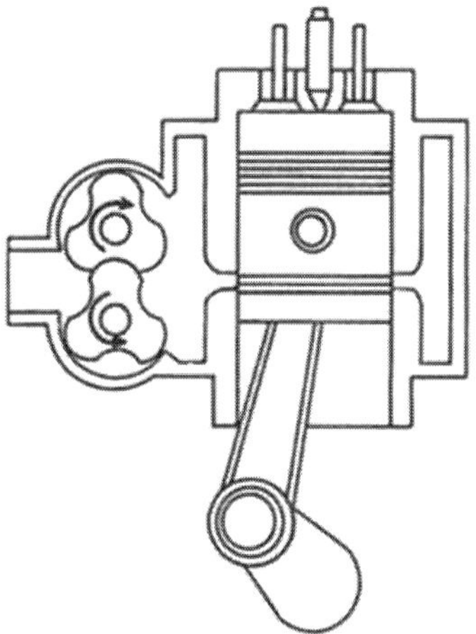

Two Cycle Diesel engine with Roots blower, typical of Detroit Diesel and someElectro-Motive Diesel Engines

There are three size groups of Diesel engines

- Small—under 188 kW (252 hp) output
- Medium
- Large

Basic types

There are two basic types of Diesel Engines

- Four stroke cycle
- Two stroke cycle

Early engines

Rudolf Diesel based his engine on Nikolaus Otto's 1876 engine, with the goal of improving its efficiency. Diesel's engine concepts were set forth in patents in 1892 and 1893. As such, diesel engines in the late 19th and early 20th centuries used the same basic layout and form as industrial steam engines, with long-bore cylinders, external valve gear, cross-head bearings and an open crankshaft connected to a large flywheel. Smaller engines would be built with vertical cylinders, while most medium- and large-sized industrial engines were built with horizontal cylinders, just as steam engines had been. Engines could be built with more than one cylinder in both cases. The largest early diesels resembled the triple-expansion steam reciprocating engine, being tens of feet high with vertical cylinders arranged in-line. These early engines ran at very slow speeds—partly due to the limitations of their air-blast injector equipment and partly so they would be compatible with the majority of industrial equipment designed for steam engines; maximum speeds of 100–300 rpm were common. Engines were usually started by allowing compressed air into the cylinders to turn the engine, although smaller engines could be started by hand.

In 1897, when the first Diesel engine was completed Adolphus Busch traveled to Cologne and negotiated exclusive right to produce the Diesel engine in the USA and Canada. In his examination of the engine, it was noted that the Diesel at that time operated at thermodynamic efficiencies of 32–35%, while a typical triple expansion steam engine would operate at about 18%.

In the early decades of the 20th century, when large diesel engines were first being used, the engines took a form similar to the compound steam engines common at the time, with the piston being connected to the connecting rod by a crosshead bearing. Following steam engine practice some manufacturers made double-acting two-stroke and four-stroke diesel engines to increase power output, with combustion taking place on both sides of the piston, with two sets of valve gear and fuel injection. While it produced large amounts of power and was very efficient, the double-acting diesel engine's main problem was producing a good seal where the piston rod passed through the bottom of the lower combustion chamber to the crosshead bearing, and no more were built. By the 1930s turbochargers were fitted to some engines. Crosshead bearings are still used to reduce the wear on the cylinders in large long-stroke main marine engines.

Modern high and medium-speed engines

A Yanmar 2GM20 marine diesel engine, installed in a sailboat

As with petrol engines, there are two classes of diesel engines in current use: two-stroke and four-stroke. The four-stroke type is the "classic" version, tracing its lineage back to Rudolf Diesel's prototype. It is also the most commonly used form, being the preferred power source for many motor vehicles, especially buses and trucks. Much larger engines, such as used for railroad locomotion and marine propulsion, are often two-stroke units, offering a more favourable power-to-weight ratio, as well as better fuel economy. The most powerful engines in the world are two-stroke diesels of mammoth dimensions.

Two-stroke diesel engine operation is similar to that of petrol counterparts, except that fuel is not mixed with air before induction, and the crankcase does not take an active role in the cycle. The traditional two-stroke design relies upon a mechanically driven positive displacement blower to charge the cylinders with air before compression and ignition. The charging process also assists in expelling (scavenging) combustion gases remaining from the previous power stroke.

The archetype of the modern form of the two-stroke diesel is the (high-speed) Detroit Diesel Series 71 engine, designed by Charles F. "Boss" Kettering and his colleagues at General Motors Corporation in 1938, in which the blower pressurizes a chamber in the engine block that is often rcfcrred to as the "air box". The (very much larger medium-speed) Electro-Motive Diesel engine is used as theprime mover in EMD diesel-electric locomotive, marine and stationary applications, and was designed by the same team, and is built to the same principle. However, a significant improvement built into most later EMD engines is the mechanically assisted turbo-compressor, which provides charge air using mechanical assistance during starting (thereby obviating the necessity for Roots-blown scavenging), and provides charge air using an exhaust gas-driven turbine during normal operations—thereby providing true turbocharging and additionally increasing the engine's power output by at least fifty percent.

Three English Electric 7SRL diesel-alternator sets being installed at the Saateni Power Station, Zanzibar 1955

In a two-stroke diesel engine, as the cylinder's piston approaches the bottom dead centre exhaust ports or valves are opened relieving most of the excess pressure after which a passage between the air box and the cylinder is opened, permitting air flow into the cylinder.The air flow blows the remaining combustion gases from the cylinder—this is the scavenging process. As the piston passes through bottom centre and starts upward, the passage is closed and compression commences, culminating in fuel injection and ignition.Refer to two-stroke diesel engines for more detailed coverage of aspiration types and supercharging of two-stroke diesel engines.

Normally, the number of cylinders are used in multiples of two, although any number of cylinders can be used as long as the load on the crankshaft is counterbalanced to prevent excessive vibration. The inline-six-cylinder design is the most prolific in light- to medium-duty engines, though small V8 and larger inline-four displacement engines are also common. Small-capacity engines (generally considered to be those below five litres in capacity) are generally four- or six-cylinder types, with the four-cylinder being the most common type found in automotive uses. Five-cylinder diesel engines have also been produced, being a compromise between the smooth running of the six-cylinder and the space-efficient dimensions of the four-cylinder. Diesel engines for smaller plant machinery, boats, tractors, generators and pumps may be four, three or two-cylinder types, with the single-cylinder diesel engine remaining for light stationary work. Direct reversible two-stroke marine diesels need at least three cylinders for reliable restarting forwards and reverse, while four-stroke diesels need at least six cylinders. The desire to improve the diesel engine's power-to-weight ratio produced several novel cylinder arrangements to extract more power from a given capacity. The uniflowopposed-piston engine uses two pistons in one cylinder with the combustion cavity in the middle and gas in- and outlets at the ends. This makes a comparatively light, powerful, swiftly running and economic engine suitable for use in aviation. An example is the Junkers Jumo 204/205. The Napier Deltic engine, with three cylinders arranged in a triangular formation, each containing two opposed pistons, the whole engine having three crankshafts, is one of the better known.

MODERN LOW-SPEED ENGINES

Low-speed diesel engines (as used in ships and other applications where overall engine weight is relatively unimportant) often have a thermal efficiency which exceeds 50%.

Gas generator

Before 1950, Sulzer started experimenting with two-stroke engines with boost pressures as high as 6 atmospheres, in which all the output power was taken from an exhaust gas turbine. The two-stroke pistons directly drove air compressor pistons to make a positive displacement gas generator. Opposed pistons were connected by linkages instead of crankshafts. Several of these units could be connected to provide power gas to one large output turbine. The overall thermal efficiency was roughly twice that of a simple gas turbine. This system was derived from Raúl Pateras Pescara's work on free-piston engines in the 1930s.

ADVANTAGES AND DISADVANTAGES VERSUS SPARK-IGNITION ENGINES

Fuel economy

The MAN S80ME-C7 low speed diesel engines use 155 grams (5.5 oz) of fuel per kWh for an overall energy conversion efficiency of 54.4%, which is the highest conversion of fuel into power by any single-cycle internal or external combustion engine (The efficiency of a combined cycle gas turbine system can exceed 60%.)

Diesel engines are more efficient than gasoline (petrol) engines of the same power rating, resulting in lower fuel consumption. A common margin is 40% more miles per gallon for an efficientturbodiesel. For example, the current model Škoda Octavia, using Volkswagen Group engines, has a combined Euro rating of 6.2 L/100 km (46 mpg_{-imp}; 38 mpg_{-US}) for the 102 bhp (76 kW) petrol engine and 4.4 L/100 km (64 mpg_{-imp}; 53 mpg_{-US}) for the 105 bhp (78 kW) diesel engine.

However, such a comparison does not take into account that diesel fuel is denser and contains about 15% more energy by volume. Although the calorific value of the fuel is slightly lower at 45.3 MJ/kg (megajoules per kilogram) than petrol at 45.8 MJ/kg, liquid diesel fuel is significantly denser than liquid petrol. This is significant because volume of fuel, in addition to mass, is an important consideration in mobile applications.

Adjusting the numbers to account for the energy density of diesel fuel, the overall energy efficiency is still about 20% greater for the diesel version.

While a higher compression ratio is helpful in raising efficiency, diesel engines are much more efficient than gasoline (petrol) engines when at low power and at engine idle. Unlike the petrol engine, diesels lack a butterfly valve (throttle) in the inlet system, which closes at idle. This creates parasitic loss and destruction of availability of the incoming air, reducing the efficiency of petrol engines at idle. In many applications, such as marine, agriculture, and railways, diesels are left idling and unattended for many hours, sometimes even days. These advantages are especially attractive in locomotives.

Even though diesel engines have a theoretical fuel efficiency of 75%, in practice it is lower. Engines in large diesel trucks, buses, and newer diesel cars can achieve peak efficiencies around 45%, and could reach 55% efficiency in the near future. However, average efficiency over a driving cycle is lower than peak efficiency. For example, it might be 37% for an engine with a peak efficiency of 44%.

Torque

Diesel engines produce more torque than petrol engines for a given displacement due to their higher compression ratio. Higher pressure in the cylinder and higher forces on the connecting rods and crankshaft require stronger, heavier components. Heavier rotating components prevent diesel engines from reving as high as petrol engines for a given displacement. Diesel engines generally have similar power and inferior power to weight to petrol engines.

Petrol engines must be geared lower to get the same torque as a comparable diesel but since petrol engines rev higher both will have similar acceleration. An arbitrary amount of torque at the wheels can be gained by gearing any power source down sufficiently (including a hand crank). For example, a theoretical engine with a constant 200 ft/lbs of torque and a 3000 rpm rev limit has just as much power (a little over 114 hp) as another theoretical engine with a constant maximum 100 ft/lbs of torque and a 6000 rpm rev limit. A (lossless) 2 to 1 reduction gear on the second engine will output a constant maximum 200 ft/lbs of torque at a maximum of 3000 rpm, with no change in power. Comparing engines based on (maximum) torque is just as useful as comparing them based on (maximum) rpm.

Power

In diesel engines, conditions in the engine differ from the spark-ignition engine, since power is directly controlled by the fuel supply, rather than by controlling the air supply.

The average diesel engine has a poorer power-to-weight ratio than the petrol engine. This is because the diesel must operate at lower engine speeds and because it needs heavier, stronger parts to resist the operating pressure caused by the high compression ratio of the engine and the large amounts of torque generated to the crankshaft. In addition, diesels are often built with stronger parts to give them longer lives and better reliability, important considerations in industrial applications.

Diesel engines usually have longer stroke lengths chiefly to facilitate achieving the necessary compression ratios, but also to reduce the optimal operating speed (rpm). As a result, piston and connecting rods are heavier and more force must be transmitted through the connecting rods and crankshaft to change the momentum of the piston. This is another reason that a diesel engine must be stronger for the same power output as a petrol engine.

Yet it is this characteristic that has allowed some enthusiasts to acquire significant power increases with turbocharged engines by making fairly simple and inexpensive modifications. A petrol engine of similar size cannot put out a comparable power increase without extensive alterations because the stock components cannot withstand the higher stresses placed upon them. Since a diesel engine is already built to withstand higher levels of stress, it makes an ideal candidate for performance tuning at little expense. However, it should be said that any modification that raises the amount of fuel and air put through a diesel engine will increase its operating temperature, which will reduce its life and increase service requirements. These are issues with newer, lighter, *high-performance* diesel engines which are not "overbuilt" to the degree of older engines and they are being pushed to provide greater power in smaller engines.

Forced induction

The addition of a turbocharger or supercharger to the engine does not assist in increasing fuel economy, but will increase power output in the same sized engine, mitigating the fuel-air intake speed limit mentioned above for a given engine displacement. Boost pressures can be higher on diesels than on petrol engines, due to the latter's susceptibility to knock, and the higher compression ratio allows a diesel engine to be more efficient than a comparable spark ignition engine. Because the burned gases are expanded further in a diesel engine cylinder, the exhaust gas is cooler, meaning turbochargers require less cooling, and can be more reliable, than with spark-ignition engines.

Without the risk of knocking, boost pressure in a diesel engine can be much higher; it is possible to run as much boost as the engine will physically stand before breaking apart. A combination of improved mechanical technology (such as multi-stage injectors which fire a short "pilot charge" of fuel into the cylinder to warm the combustion chamber before delivering the main fuel charge), higher injection pressures that have improved the atomisation of fuel into smaller droplets, and electronic control (which can adjust the timing and length of the injection process to optimise it for all speeds and temperatures) have mitigated most of these problems in the latest generation of common-rail designs, while greatly improving engine efficiency. Poor power and narrow torque bands have been addressed by superchargers, turbochargers, (especially variable geometry turbochargers),intercoolers, and a large efficiency increase from about 35% for IDI to 45% for the latest engines in the last 15 years.

Emissions

Since the diesel engine uses less fuel than the petrol engine per unit distance, the diesel produces less carbon dioxide (CO_2) per unit distance. Recent advances in production and changes in the political climate have increased the availability and awareness of biodiesel, an alternative to petroleum-derived diesel fuel with a much lower net-sum emission of CO_2, due to the absorption of CO_2 by plants used to produce the fuel. Although concerns are now being raised as to the negative effect this is having on the world food supply, as the growing of crops specifically for biofuels takes up land that could be used for food crops and uses water that could be used by both humans and animals. However, the use of waste vegetable oil, sawmill waste from managed forests in Finland, and advances in the production of vegetable oil from algae demonstrate great promise in providing feed stocks for sustainable biodiesel that are not in competition with food production.

When a diesel engine runs at low power, there is enough oxygen present to burn the fuel—diesel engines only make significant amounts of carbon monoxide when running under a load.

Diesel fuel is injected just before the power stroke. As a result, the fuel cannot burn completely unless it has a sufficient amount of oxygen. This can result in incomplete combustion and black smoke in the exhaust if more fuel is injected than there is air available for the combustion process. Modern engines with electronic fuel delivery can adjust the timing and amount of fuel delivery (by changing the duration of the injection pulse), and so operate with less waste of fuel. In a mechanical system, the injection timing and duration must be set to be efficient at the anticipated operating rpm and load, and so the settings are less than ideal when the engine is running at any other RPM than what it is timed for. The electronic injection can "sense" engine revs, load, even boost and temperature, and continuously alter the timing to match the given situation. In the petrol engine, air and fuel are mixed for the entire compression stroke, ensuring complete mixing even at higher engine speeds.

Diesel exhaust is well known for its characteristic smell; but this smell in recent years has become much less because the sulfur is now removed from the fuel in the oil refinery.

Diesel exhaust has been found to contain a long list of toxic air contaminants. Among these pollutants, fine particle pollution is an important as a cause of diesel's harmful health effects. However, when diesel engines burn their fuel with high oxygen levels, this results in high combustion temperatures and higher efficiency, and these particles tend to burn off and reduce, but the amount of NOx pollution tends to increase.

NOx pollution can be reduced with diesel exhaust fluid, which is injected into the exhaust stream, and catalytically destroys the NOx chemical species. Exhaust gas recirculationwhich works by recirculating a portion of an engine's exhaust gas back to the engine cylinders also has very positive effects on NOx emissions.

Noise

The distinctive noise of a diesel engine is variably called diesel clatter, diesel nailing, or diesel knock. Diesel clatter is caused largely by the diesel combustion process; the sudden ignition of the diesel fuel when injected into the combustion chamber causes a pressure wave. Engine designers can reduce diesel clatter through: indirect injection; pilot or pre-injection; injection timing; injection rate; compression ratio; turbo boost; and exhaust gas recirculation (EGR). Common rail diesel injection systems permit multiple injection events as an aid to noise reduction. Diesel fuels with a higher cetane rating modify the combustion process and reduce diesel clatter. CN (Cetane number) can be raised by distilling higher quality crude oil, by catalyzing a higher quality product or by using a cetane improving additive.

A combination of improved mechanical technology such as multi-stage injectors which fire a short "pilot charge" of fuel into the cylinder to initiate combustion before delivering the main fuel charge, higher injection pressures that have improved the atomisation of fuel into smaller droplets, and electronic control (which can adjust the timing and length of the injection process to optimise it for all speeds and temperatures), have partially mitigated these problems in the latest generation of common-rail designs, while improving engine efficiency.

Reliability

For most industrial or nautical applications, reliability is considered more important than light weight and high power.

The lack of an electrical ignition system greatly improves the reliability. The high durability of a diesel engine is also due to its overbuilt nature, a benefit that is magnified by the lower rotating speeds in diesels. Diesel fuel is a better lubricant than petrol and thus, it is less harmful to the oil film on piston rings and cylinder bores; it is routine for diesel engines to cover 400,000 km (250,000 mi) or more without a rebuild.

Due to the greater compression ratio and the increased weight of the stronger components, starting a diesel engine is harder than starting a gasoline engine of similar design and displacement. More torque from the starter motor is required to push the engine through the compression cycle when starting compared to a petrol engine. This can cause difficulty when starting in winter time if using conventional automotive batteries because of the lower current available.

Either an electrical starter or an air-start system is used to start the engine turning. On large engines, pre-lubrication and slow turning of an engine, as well as heating, are required to minimise the amount of engine damage during initial start-up and running. Some smaller military diesels can be started with an explosive cartridge, called a Coffman starter, which provides the extra power required to get the machine turning. In the past, Caterpillar and John Deere used a small petrol *pony* engine in their tractors to start the primary diesel engine. The pony engine heated the diesel to aid in ignition and used a small clutch and transmission to spin up the diesel engine. Even more unusual was anInternational Harvester design in which the diesel engine had its own carburetor and ignition system, and started on petrol. Once warmed up, the operator moved two levers to switch the engine to diesel operation, and work could begin. These engines had very complex cylinder heads, with their own petrol combustion chambers, and were vulnerable to expensive damage if special care was not taken (especially in letting the engine cool before turning it off).

CYLINDER CAVITATION AND EROSION DAMAGE

One phenomenon that can affect water-cooled diesel engines is cylinder cavitation and erosion. This is due to a phenomenon in high-compression engines where the ignition of the fuel in the cylinder causes a high-frequency vibration that causes bubbles to form in the coolant in contact with the cylinder. When these tiny bubbles collapse, coolant impacts the cylinder wall, over time causing small holes to form in the cylinder wall. This damage is mitigated in some engines with coatings, or with a coolant additive specifically designed to prevent cavitation and erosion damage. Engines damaged in this way will require the affected cylinder to be repaired (where possible) or will be rendered unusable.

Quality and variety of fuels

Petrol/gasoline engines are limited in the variety and quality of the fuels they can burn. Older petrol engines fitted with a carburetor required a volatile fuel that would vaporise easily to create the necessary air-fuel ratio for combustion. Because both air and fuel are admitted to the cylinder, if the compression ratio of the engine is too high or the fuel too volatile (with too low an octane rating), the fuel will ignite under compression, as in a diesel engine, before the piston reaches the top of its stroke. This pre-ignition causes a power loss and over time major damage to the piston and cylinder. The need for a fuel that is volatile enough to vaporise but not too volatile (to avoid pre-ignition) means that petrol engines will only run on a narrow range of fuels. There has been some success at dual-fuel engines that use petrol and ethanol, petrol and propane, and petrol andmethane.

In diesel engines, a mechanical injector system vaporizes the fuel directly into the combustion chamber or a pre-combustion chamber (as opposed to a Venturi jet in a carburetor, or a fuel injector in a fuel injection system vaporising fuel into the intake manifold or intake runners as in a petrol engine). This*forced vaporisation* means that less-volatile fuels can be used. More crucially, because only air is inducted into the cylinder in a diesel engine, the compression ratio can be much higher as there is no risk of pre-ignition provided the injection process is accurately timed. This means that cylinder temperatures are much higher in a diesel engine than a petrol engine, allowing less volatile fuels to be used.

Diesel fuel is a form of light fuel oil, very similar to kerosene (paraffin), but diesel engines, especially older or simple designs that lack precision electronic injection systems, can run on a wide variety of other fuels. Some of the most common alternatives are Jet A-1 type jet fuel or vegetable oil from a very wide variety of plants. Some engines can be run on vegetable oil without modification, and most others require fairly basic alterations. Biodiesel is a pure diesel-like fuel refined from vegetable oil and can be used in nearly all diesel engines. Requirements for fuels to be used in diesel engines are the ability of the fuel to flow along the fuel lines, the ability of the fuel to lubricate the injector pump and injectors adequately, and its ignition qualities (ignition delay, cetane number). Inline mechanical injector pumps generally tolerate poor-quality or bio-fuels better than distributor-type pumps. Also, indirect injection engines generally run more satisfactorily on bio-fuels than direct injection engines. This is partly because an indirect injection engine has a much greater 'swirl' effect, improving vaporisation and combustion of fuel, and because (in the case of vegetable oil-type fuels) lipid depositions can condense on the cylinder walls of a direct-injection engine if combustion temperatures are too low (such as starting the engine from cold).

It is often reported that Diesel designed his engine to run on peanut oil, but this is false. Patent number 608845 describes his engine as being designed to run on pulverulent solid fuel (coal dust). Diesel stated in his published papers, "at the Paris Exhibition in 1900 (*Exposition Universelle*) there was shown by the Otto Company a small diesel engine, which, at the request of the French

Government ran on Arachide (earth-nut or peanut) oil, and worked so smoothly that only a few people were aware of it. The engine was constructed for using mineral oil, and was then worked on vegetable oil without any alterations being made. The French Government at the time thought of testing the applicability to power production of the Arachide, or earth-nut, which grows in considerable quantities in their African colonies, and can easily be cultivated there." Diesel himself later conducted related tests and appeared supportive of the idea.

Most large marine diesels run on heavy fuel oil (sometimes called "bunker oil"), which is a thick, viscous and almost flameproof fuel which is very safe to store and cheap to buy in bulk as it is a waste product from the petroleum refining industry. The fuel must not only be pre-heated, but must be kept heated during handling and storage in order to maintain its pumpability. This is usually accomplished by steam tracing on fuel lines and steam coils in fuel oil tanks. The fuel is then preheated to over 100C before entering the engine in order to attain the proper viscosity for atomisation.

FUEL AND FLUID CHARACTERISTICS

Diesel engines can operate on a variety of different fuels, depending on configuration, though the eponymous diesel fuel derived from crude oil is most common. The engines can work with the full spectrum of crude oil distillates, from natural gas, alcohols, petrol, wood gas to the *fuel oils* from diesel oil to residual fuels. Many automotive diesel engines would run on 100% biodiesel without any modifications. This would be such a potential advantage since biodiesel can be made so much more cheaply than it takes to have traditional diesel fuel from a fuel station's pump.

The type of fuel used is selected to meet a combination of service requirements, and fuel costs. Good-quality diesel fuel can be synthesised from vegetable oil and alcohol. Diesel fuel can be made from coal or other carbon base using the Fischer–Tropsch process. Biodiesel is growing in popularity since it can frequently be used in unmodified engines, though production remains limited. Recently, biodiesel from coconut, which can produce a very promising coco methyl ester (CME), has characteristics which enhance lubricity and combustion giving a regular diesel engine without any modification more power, less particulate matter or black smoke, and smoother engine performance. The Philippines pioneers in the research on Coconut based CME with the help of German and American scientists. Petroleum-derived diesel is often called *petrodiesel* if there is need to distinguish the source of the fuel.

Pure plant oils are increasingly being used as a fuel for cars, trucks and remote combined heat and power generation especially in Germany where hundreds of decentralised small- and medium-sized oil presses cold press oilseed, mainly rapeseed, for fuel. There is a Deutsches Institut für Normung fuel standard for rapeseed oil fuel.

Residual fuels are the "dregs" of the distillation process and are a thicker, heavier oil, or oil with higher viscosity, which are so thick that they are not readily pumpable unless heated. Residual fuel oils are cheaper than clean, refined diesel oil, although they are dirtier. Their main considerations are for use in ships and very large generation sets, due to the cost of the large volume of fuel consumed, frequently amounting to many tonnes per hour. The poorly refined biofuels straight vegetable oil (SVO) and waste vegetable oil(WVO) can fall into this category, but can be viable fuels on non-common rail or TDI PD diesels with the simple conversion of fuel heating to 80 to 100 degrees Celsius to reduce viscosity, and adequate filtration to OEM standards. Engines using these heavy oils have to start and shut down on standard diesel fuel, as these fuels will not flow through fuel lines at low temperatures. Moving beyond that, use of low-grade fuels can lead to serious maintenance problems because of their high sulphur and lower lubrication properties. Most diesel engines that power ships like supertankers are built so that the engine can safely use low-grade fuels due to their separate cylinder and crankcase lubrication.

Normal diesel fuel is more difficult to ignite and slower in developing fire than petrol because of its higher flash point, but once burning, a diesel fire can be fierce.

Fuel contaminants such as dirt and water are often more problematic in diesel engines than in petrol engines. Water can cause serious damage, due to corrosion, to the injection pump and injectors; and dirt, even very fine particulate matter, can damage the injection pumps due to the close tolerances that the pumps are machined to. All diesel engines will have a fuel filter (usually much finer than a filter on a petrol engine), and a water trap. The water trap (which is sometimes part of the fuel filter) often has a float connected to a warning light, which warns when there is too much water in the trap, and must be drained before damage to the engine can result. The fuel filter must be replaced much more often on a diesel engine than on a petrol engine, changing the fuel filter every 2–4 oil changes is not uncommon for some vehicles.

SAFETY

Fuel flammability

Diesel fuel has low flammability, leading to a low risk of fire caused by fuel in a vehicle equipped with a diesel engine.

In yachts, diesel engines are often used because the petrol (gasoline) that fuels spark-ignition engines releases combustible vapors which can lead to an explosion if it accumulates in a confined space such as the bottom of a vessel. Ventilation systems are mandatory on petrol-powered vessels.

The United States Army and NATO use only diesel engines and turbines because of fire hazard. Although neither gasoline nor diesel is explosive in liquid form, both can create an explosive air/vapor mix under the right conditions. However, diesel fuel is less prone due to its lower vapor pressure, which is an indication of evaporation rate. The Material Safety Data Sheet for ultra-low sulfur diesel fuel indicates a vapor explosion hazard for diesel indoors, outdoors, or in sewers.

US Army gasoline-engined tanks during World War II were nicknamed Ronsons, because of their greater likelihood of catching fire when damaged by enemy fire. (Although tank fires were usually caused by detonation of the ammunition rather than fuel), while diesel tanks such as the Soviet T-34 were less prone to catching fire.

Maintenance hazards

Fuel injection introduces potential hazards in engine maintenance due to the high fuel pressures used. Residual pressure can remain in the fuel lines long after an injection-equipped engine has been shut down. This residual pressure must be relieved, and if it is done so by external bleed-off, the fuel must be safely contained. If a high-pressure diesel fuel injector is removed from its seat and operated in open air, there is a risk to the operator of injury by hypodermic jet-injection, even with only 100 pounds per square inch (690 kPa) pressure. The first known such injury occurred in 1937 during a diesel engine maintenance operation.

Cancer

Diesel exhaust has been classified as an IARC Group 1 carcinogen. It causes lung cancer and is associated with an increased risk for bladder cancer.

APPLICATIONS

The characteristics of diesel have different advantages for different applications.

Passenger cars

Diesel engines have long been popular in bigger cars and have been used in smaller cars such as superminis like the Peugeot 205, in Europe since the 1980s. Diesel engines tend to be more economical at regular driving speeds and are much better at city speeds. Their reliability and life-span tend to be better (as detailed). Some 40% or more of all cars sold in Europe are diesel-powered where they are considered a low CO_2 option. Mercedes-Benz in conjunction with Robert Bosch GmbH produced diesel-powered passenger cars starting in 1936 and very large numbers are used all over the world (often as "Grande Taxis" in the Third World). Diesel-powered passenger cars are very popular in India too, since the price of diesel fuel there is lower as compared to petrol. As a result, predominantly petrol-powered car manufacturers including the Japanese car manufacturers produce and market diesel-powered cars in India. Diesel-powered cars also dominate the Indian taxi industry.

Railroad rolling stock

Diesel engines have eclipsed steam engines as the prime mover on all non-electrified railroads in the industrialized world. The first diesel locomotives appeared in the early 20th century, and diesel multiple units soon after. While electric locomotives have replaced the diesel locomotive for some passenger traffic in Europe and Asia, diesel is still today very popular for cargo-hauling freight trains and on tracks where electrification is not feasible. Most modern diesel locomotives are actually diesel-electric locomotives: the diesel engine is used to power an electric generator that in turn powers electric traction motors with no mechanical connection between diesel engine and traction. After 2000, environmental requirements has caused higher development cost for engines, and it has become common for passenger multiple units to use engines and automatic mechanical gearboxes made for trucks. Up to four such combinations might be used to get enough power in a train.

Other transport uses

Larger transport applications (trucks, buses, etc.) also benefit from the Diesel's reliability and high torque output. Diesel displaced paraffin (or tractor vaporising oil, TVO) in most parts of the world by the end of the 1950s with the U.S. following some 20 years later.

- Aircraft
- Marine
- Motorcycles

In merchant ships and boats, the same advantages apply with the relative safety of Diesel fuel an additional benefit. The German pocket battleships were the largest Diesel warships, but the German torpedo-boats known as E-boats (*Schnellboot*) of the Second World War were also Diesel craft. Conventional submarines have used them since before World War I, relying on the almost total absence of carbon monoxide in the exhaust. American World War II Diesel-electric submarines operated on two-stroke cycle, as opposed to the four-stroke cycle that other navies used.

Non-road diesel engines

Non-road diesel engines include mobile equipment and vehicles that are not used on the public roadways such as construction equipment and agricultural tractors.

Military fuel standardisation

NATO has a single vehicle fuel policy and has selected diesel for this purpose. The use of a single fuel simplifies wartime logistics. NATO and the

United States Marine Corps have even been developing a diesel military motorcycle based on a Kawasaki off road motorcycle the KLR 650, with a purpose designed naturally aspirated direct injection diesel at Cranfield University in England, to be produced in the USA, because motorcycles were the last remaining gasoline-powered vehicle in their inventory. Before this, a few civilianmotorcycles had been built using adapted stationary diesel engines, but the weight and cost disadvantages generally outweighed the efficiency gains.

Non-transport uses

Diesel engines are also used to power permanent, portable, and backup generators, irrigation pumps, corn grinders, and coffee de-pulpers.

Engine speeds

Within the diesel engine industry, engines are often categorized by their rotational speeds into three unofficial groups:

- High-speed engines (> 1,000 rpm),
- Medium-speed engines (300–1,000 rpm), and
- Slow-speed engines (< 300 rpm).

High- and medium-speed engines are predominantly four-stroke engines; except for the Detroit Diesel two-stroke range. Medium-speed engines are physically larger than high-speed engines and can burn lower-grade (slower-burning) fuel than high-speed engines. Slow-speed engines are predominantly large two-stroke crosshead engines, hence very different from high- and medium-speed engines. Due to the lower rotational speed of slow- and medium-speed engines, there is more time for combustion during the power stroke of the cycle, allowing the use of slower-burning fuels than high-speed engines.

High-speed engines

High-speed (approximately 1,000 rpm and greater) engines are used to power trucks (lorries), buses, tractors, cars, yachts, compressors, pumps and small electrical generators. As of 2008, most high-speed engines have direct injection. Many modern engines, particularly in on-highway applications, have common rail direct injection, which is cleaner burning.

Medium-speed engines

Medium-speed engines are used in large electrical generators, ship propulsion and mechanical drive applications such as large compressors or pumps. Medium speed diesel engines operate on either diesel fuel or heavy fuel oil by direct injection in the same manner as low-speed engines.

Engines used in electrical generators run at approximately 300 to 1000 rpm and are optimized to run at a set synchronous speed depending on the generation frequency (50 or 60 hertz) and provide a rapid response to load changes. Typical synchronous speeds for modern medium-speed engines are 500/514 rpm (50/60 Hz), 600 rpm (both 50 and 60 Hz), 720/750 rpm, and 900/1000 rpm.

As of 2009, the largest medium-speed engines in current production have outputs up to approximately 20 MW (27,000 hp) and are supplied by companies like MAN B&W,Wärtsilä, and Rolls-Royce (who acquired Ulstein Bergen Diesel in 1999). Most medium-speed engines produced are four-stroke machines, however there are some two-stroke medium-speed engines such as by EMD (Electro-Motive Diesel), and the Fairbanks Morse OP (Opposed-piston engine) type.

Typical cylinder bore size for medium-speed engines ranges from 20 cm to 50 cm, and engine configurations typically are offered ranging from in-line 4-cylinder units to V-configuration 20-cylinder units. Most larger medium-speed engines are started with compressed air direct on pistons, using an air distributor, as opposed to a pneumatic starting motor acting on the flywheel, which tends to be used for smaller engines. There is no definitive engine size cut-off point for this.

It should also be noted that most major manufacturers of medium-speed engines make natural gas-fueled versions of their diesel engines, which in fact operate on the Otto cycle, and require spark ignition, typically provided with a spark plug. There are also dual (diesel/natural gas/coal gas) fuel versions of medium and low speed diesel engines using a lean fuel air mixture and a small injection of diesel fuel (so-called "pilot fuel") for ignition. In case of a gas supply failure or maximum power demand these engines will instantly switch back to full diesel fuel operation.

Low-speed engines

The MAN B&W 5S50MC 5-cylinder, 2-stroke, low-speed marine diesel engine. This particular engine is found aboard a 29,000 tonne chemical carrier.

Also known as *slow-speed*, or traditionally *oil engines*, the largest diesel engines are primarily used to power ships, although there are a few land-based power generation units as well. These extremely large two-stroke engines have power outputs up to approximately 85 MW (114,000 hp), operate in the range from approximately 60 to 200 rpm and are up to 15 m (50 ft) tall, and can weigh over 2,000 short tons (1,800 t). They typically use direct injection running on cheap low-grade heavy fuel, also known as bunker C fuel, which requires heating in the ship for tanking and before injection due to the fuel's high viscosity. Often, the waste heat recovery steam boilers attached to the engine exhaust ducting generate the heat required for fuel heating. Provided the heavy fuel system is kept warm and circulating, engines can be started and stopped on heavy fuel.

Large and medium marine engines are started with compressed air directly applied to the pistons. Air is applied to cylinders to start the engine forwards or backwards because they are normally directly connected to the propeller without clutch or gearbox, and to provide reverse propulsion either the engine must be run backwards or the ship will use an adjustable propeller. At least three cylinders are required with two-stroke engines and at least six cylinders with four-stroke engines to provide torque every 120 degrees.

Companies such as MAN B&W Diesel, (formerly Burmeister & Wain) and Wärtsilä (which acquired Sulzer Diesel) design such large low-speed engines. They are unusually narrow and tall due to the addition of a crosshead bearing. As of 2007, the 14-cylinder Wärtsilä-Sulzer 14RTFLEX96-C turbocharged two-stroke diesel engine built by Wärtsilä licensee Doosan in Korea is the most powerful diesel engine put into service, with a cylinder bore of 960 mm (37.8 in) delivering 114,800 hp (85.6 MW).

It was put into service in September 2006, aboard what was then the world's largest container ship *Emma Maersk* which belongs to the A.P. Moller-Maersk Group. Typical bore size for low-speed engines ranges from approximately 35 to 98 cm (14 to 39 in). As of 2008, all produced low-speed engines with crosshead bearings are in-line configurations; no Vee versions have been produced.

CURRENT AND FUTURE DEVELOPMENTS

As of 2008, many common rail and unit injection systems already employ new injectors using stacked piezoelectric wafers in lieu of a solenoid, giving finer control of the injection event.

Variable geometry turbochargers have flexible vanes, which move and let more air into the engine depending on load. This technology increases both performance and fuel economy. Boost lag is reduced as turbo impeller inertia is compensated for.

Accelerometer pilot control (APC) uses an accelerometer to provide feedback on the engine's level of noise and vibration and thus instruct the ECU to inject the minimum amount of fuel that will produce quiet combustion and still provide the required power (especially while idling).

The next generation of common rail diesels is expected to use variable injection geometry, which allows the amount of fuel injected to be varied over a wider range, and variable valve timing similar to that of petrol engines. Particularly in the United States, coming tougher emissions regulations present a considerable challenge to diesel engine manufacturers. Ford's HyTrans Project has developed a system which starts the ignition in 400 ms, saving a significant amount of fuel on city routes, and there are other methods to achieve even more efficient combustion, such as homogeneous charge compression ignition, being studied.

Japanese and Swedish vehicle manufacturers are also developing diesel engines that run on dimethyl ether (DME).

Some recent diesel engine models utilize a copper alloy heat exchanger technology (CuproBraze) to take advantage of benefits in terms of thermal performance, heat transfer efficiency, strength/durability, corrosion resistance, and reduced emissions from higher operating temperatures.

Low heat rejection engines

A special class of experimental prototype internal combustion piston engines has been developed over several decades with the goal of improving efficiency by reducing heat loss. These engines are variously called adiabatic engines; due to better approximation of adiabatic expansion; low heat rejection engines, or high temperature engines.They are generally piston engines with combustion chamber parts lined with ceramic thermal barrier coatings. Some make used of pistons and other parts made of titanium which has a low thermal conductivity and density. Some designs are able to eliminate the use of a cooling system and associated parasitic losses altogether. Developing lubricants able to withstand the higher temperatures involved has been a major barrier to commercialization.

MARINE PROPULSION

A view of a ship's engine room

Marine propulsion is the mechanism or system used to generate thrust to move a ship or boat across water. While paddles andsails are still used on some smaller boats, most modern ships are propelled by mechanical systems consisting of an electric motor or engine turning a propeller, or less frequently, in pump-jets, an impeller. Marine engineering is the discipline concerned with theengineering design process of marine propulsion systems.

Marine steam engines were the first mechanical engines used in marine propulsion, however they have mostly been replaced bytwo-stroke or four-stroke diesel engines, outboard motors, and gas turbine engines on faster ships. Nuclear reactors producing steam are used to propel warships and icebreakers. Nuclear reactors to power commercial vessels has not been adopted by the marine industry. Electric motors using electric battery storage have been used for propulsion on submarines and electric boats and have been proposed for energy-efficient propulsion. Development in liquefied natural gas (LNG) fueled engines are gaining recognition for their low emissions and cost advantages. Stirling engines, which are more efficient, quieter, smoother running producing less harmful emissions than diesel engines, propel a number of small submarines. The Stirling engine has yet to be upscaled for larger surface ships.

POWER SOURCES

Pre-mechanisation

Until the application of the coal-fired steam engine to ships in the early 19th century, oars or the wind were used to assist watercraft propulsion. Merchant ships predominantly used sail, but during periods when naval warfare depended on ships closing to ram or to fight hand-to-hand, galley were preferred for their manoeuvrability and speed. The Greek navies that fought in the Peloponnesian War usedtriremes, as did the Romans at the Battle of Actium. The development of naval gunnery from the 16th century onward meant that manoeuvrability took second place to broadside weight; this led to the dominance of the sail-powered warship over the following three centuries.

In modern times, human propulsion is found mainly on small boats or as auxiliary propulsion on sailboats. Human propulsion includes the push pole, rowing, and pedals. Propulsion by sail generally consists of a sail hoisted on an erect mast, supported by stays, and controlled by lines made of rope. Sails were the dominant form of commercial propulsion until the late nineteenth century, and continued to be used well into the twentieth century on routes where wind was assured and coal was not available, such as in the South American nitrate trade. Sails are now generally used for recreation and racing, although innovative applications of kites/royals, turbosails, rotorsails, wingsails, windmills and SkySails's own kite buoy-system have been used on larger modern vessels for fuel savings.

Reciprocating steam engines

SS *Ukkopekka* uses a triple expansion steam engine

The development of piston-engined steamships was a complex process. Early steamships were fueled by wood, later ones by coal or fuel oil. Early ships used stern or side paddle wheels, while later ones used screw propellers.

The first commercial success accrued to Robert Fulton's *North River Steamboat* (often called *Clermont*) in US in 1807, followed in Europeby the 45-foot *Comet* of 1812. Steam propulsion progressed considerably over the rest of the 19th century. Notable developments include the steam surface condenser, which eliminated the use of sea water in the ship's boilers. This, along with improvements in boiler technology, permitted higher steam pressures, and thus the use of higher efficiency multiple expansion (compound) engines. As the means of transmitting the engine's power, paddle wheels gave way to more efficient screw propellers.

Steam turbines

Steam turbines were fueled by coal or, later, fuel oil or nuclear power. The marine steam turbine developed by Sir Charles Algernon Parsons raised the power-to-weight ratio. He achieved publicity by demonstrating it unofficially in the 100-foot *Turbinia* at the Spithead Naval Review in 1897. This facilitated a generation of high-speed liners in the first half of the 20th century, and rendered the reciprocating steam engine obsolete; first in warships, and later in merchant vessels.

In the early 20th century, heavy fuel oil came into more general use and began to replace coal as the fuel of choice in steamships. Its great advantages were convenience, reduced manpower by removal of the need for trimmers and stokers, and reduced space needed for fuel bunkers.

In the second half of the 20th century, rising fuel costs almost led to the demise of the steam turbine. Most new ships since around 1960 have been built with diesel engines. The last major passenger ship built with steam turbines was the *Fairsky*, launched in 1984. Similarly, many steam ships were re-engined to improve fuel efficiency. One high profile example was the 1968 built *Queen Elizabeth 2* which had her steam turbines replaced with a diesel-electric propulsion plant in 1986. Most new-build ships with steam turbines are specialist vessels such as nuclear-powered vessels, and certain merchant vessels (notably Liquefied Natural Gas (LNG) and coal carriers) where the cargo can be used as bunker fuel.

LNG carriers

New LNG carriers (a high growth area of shipping) continue to be built with steam turbines. The natural gas is stored in a liquid state in cryogenic vessels aboard these ships, and a small amount of 'boil off' gas is needed to maintain the pressure and temperature inside the vessels within operating limits. The 'boil off' gas provides the fuel for the ship's boilers, which provide steam for the turbines, the simplest way to deal with the gas.

Technology to operate internal combustion engines (modified marine two-stroke diesel engines) on this gas has improved, however, such engines are starting to appear in LNG carriers; with their greater thermal efficiency, less gas is burnt. Developments have also been made in the process of re-liquifying 'boil off' gas, letting it be returned to the cryogenic tanks. The financial returns on LNG are potentially greater than the cost of the marine-grade fuel oil burnt in conventional diesel engines, so the re-liquefaction process is starting to be used on diesel engine propelled LNG carriers.

Another factor driving the change from turbines to diesel engines for LNG carriers is the shortage of steam turbine qualified seagoing engineers. With the lack of turbine powered ships in other shipping sectors, and the rapid rise in size of the worldwide LNG fleet, not enough have been trained to meet the demand. It may be that the days are numbered for marine steam turbine propulsion systems, even though all but sixteen of the orders for new LNG carriers at the end of 2004 were for steam turbine propelled ships.

The NS *Savannah* was the firstnuclear-powered cargo-passenger ship

Nuclear-powered steam turbines

In these vessels, the nuclear reactor heats water to create steam to drive the turbines. Due to low prices of diesel oil, nuclear propulsion is rare except in some Navy and specialist vessels such as icebreakers. In large aircraft carriers, the space formerly used for ship's bunkerage could be used instead to bunker aviation fuel. In submarines, the ability to run submerged at high speed and in relative quiet for long periods holds obvious advantages. A few cruisers have also employed nuclear power; as of 2006, the only ones remaining in service are the Russian *Kirov* class. An example of a non-military ship with nuclear marine propulsion is the *Arktika* class icebreaker with 75,000 shaft horsepower (55,930 kW). Commercial experiments such as the NS *Savannah* have so far proved uneconomical compared with conventional propulsion. In recent times, there is some renewed interest in commercial nuclear shipping. Nuclear-powered cargo ships could lower costs associated with carbon dioxide emissions and travel at higher cruise speeds than conventional diesel powered vessels.

Reciprocating diesel engines

A modern diesel engine aboard a cargo ship

Most modern ships use a reciprocating diesel engine as their prime mover, due to their operating simplicity, robustness and fuel economy compared to most other prime mover mechanisms. The rotating crankshaft can be directly coupled to the propeller with slow speed engines, via a reduction gearbox for medium and high speed engines, or via an alternator and electric motor in diesel-electric vessels. The rotation of the crankshaft is connected to the camshaft or a hydraulic pump on an intelligent diesel.

The reciprocating marine diesel engine first came into use in 1903 when the diesel electric rivertanker *Vandal* was put into service byBranobel. Diesel engines soon offered greater efficiency than the steam turbine, but for many years had an inferior power-to-space ratio. The advent of turbocharging however hastened their adoption, by permitting greater power densities.

Diesel engines today are broadly classified according to

- Their operating cycle: two-stroke engine or four-stroke engine
- Their construction: crosshead, trunk, or opposed piston
- Their speed
- Slow speed: any engine with a maximum operating speed up to 300 revolutions per minute (rpm), although most large two-stroke slow speed diesel engines operate below 120 rpm. Some very long stroke engines have a maximum speed of around 80 rpm. The largest, most powerful engines in the world are slow speed, two stroke, crosshead diesels.
- Medium speed: any engine with a maximum operating speed in the range 300-900 rpm. Many modern four-stroke medium speed diesel engines have a maximum operating speed of around 500 rpm.
- High speed: any engine with a maximum operating speed above 900 rpm.

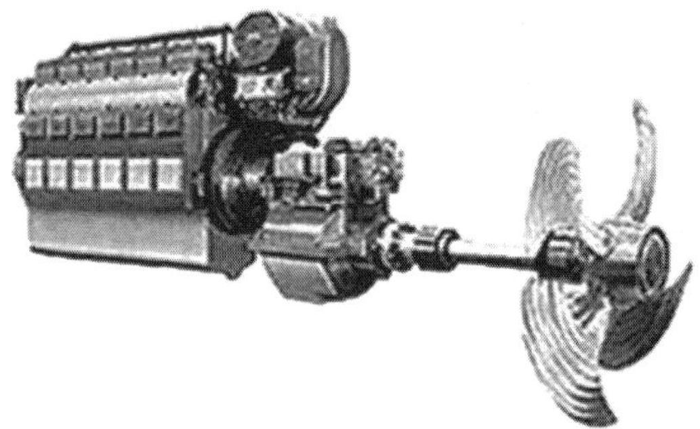

4-Stroke Marine Diesel Engine System

Most modern larger merchant ships use either slow speed, two stroke, crosshead engines, or medium speed, four stroke, trunk engines. Some smaller vessels may use high speed diesel engines.

The size of the different types of engines is an important factor in selecting what will be installed in a new ship. Slow speed two-stroke engines are much taller, but the footprint required is smaller than that needed for equivalently rated four-stroke medium speed diesel engines. As space above the waterline is at a premium in passenger ships and ferries (especially ones with a car deck), these ships tend to use multiple medium speed engines resulting in a longer, lower engine room than that needed for two-stroke diesel engines. Multiple engine installations also give redundancy in the event of mechanical failure of one or more engines, and the potential for greater efficiency over a wider range of operating conditions.

As modern ships' propellers are at their most efficient at the operating speed of most slow speed diesel engines, ships with these engines do not generally need gearboxes. Usually such propulsion systems consist of either one or two propeller shafts each with its own direct drive engine. Ships propelled by medium or high speed diesel engines may have one or two (sometimes more) propellers, commonly with one or more engines driving each propeller shaft through a gearbox. Where more than one engine is geared to a single shaft, each engine will most likely drive through a clutch, allowing engines not being used to be disconnected from the gearbox while others keep running. This arrangement lets maintenance be carried out while under way, even far from port.

LNG Engines

Shipping companies are required to comply with the International Maritime Organization (IMO) and the International Convention for the Prevention of Pollution from Shipsemissions rules. Dual fuel engines are fueled by either marine grade diesel, heavy fuel oil, or liquefied natural gas (LNG). A Marine LNG Engine has multiple fuel options, allowing vessels to transit without relying on one type of fuel. Studies show that LNG is the most efficient of fuels, although limited access to LNG fueling stations limits the production of such engines. Vessels providing services in the LNG industry have been retrofitted with dual-fuel engines, and have been proved to be extremely effective. Benefits of dual-fuel engines include fuel and operational flexibility, high efficiency, low emissions, and operational cost advantages.

Liquefied natural gas engines offer the marine transportation industry with an environmentally friendly alternative to provide power to vessels. In 2010, STX Finland and Viking Line signed an agreement to begin construction on what would be the largest environmentally friendly cruise ferry. Construction of NB 1376 will be completed in 2013. According to Viking Line, vessel NB 1376 will primarily be fueled by liquefied natural gas. Vessel NB 1376 nitrogen oxide emissions will be almost zero, and sulphur oxide emissions will be at least 80% below the International Maritime Organization's (IMO) standards. Company profits from tax cuts and operational cost advantages has led to the gradual growth of LNG fuel use in engines.

Gas turbines

Many warships built since the 1960s have used gas turbines for propulsion, as have a few passenger ships, like the jetfoil. Gas turbines are commonly used in combination with other types of engine.

Most recently, the *Queen Mary 2* has had gas turbines installed in addition to diesel engines. Because of their poor thermal efficiency at low power (cruising) output, it is common for ships using them to have diesel engines for cruising, with gas turbines reserved for when higher speeds are needed however, in the case of passenger ships the main reason for installing gas turbines has been to allow a reduction of emissions in sensitive environmental areas or while in port.

Some warships, and a few modern cruise ships have also used steam turbines to improve the efficiency of their gas turbines in a combined cycle, where waste heat from a gas turbine exhaust is utilized to boil water and create steam for driving a steam turbine.

In such combined cycles, thermal efficiency can be the same or slightly greater than that of diesel engines alone; however, the grade of fuel needed for these gas turbines is far more costly than that needed for the diesel engines, so the running costs are still higher.

Stirling engines

Since the late 1980s, Swedish shipbuilder Kockums has built a number of successful Stirling engine powered submarines. The submarines store compressed oxygen to allow more efficient and cleaner external fuel combustion when submerged, providing heat for the Stirling engine's operation. The engines are currently used on submarines of the *Gotland* and *Södermanland* classes. and the Japanese *Sôryû*-class submarine. These are the first submarines to feature Stirling air-independent propulsion (AIP), which extends the underwater endurance from a few days to several weeks.

The heat sink of a Stirling engine is typically the ambient air temperature. In the case of medium to high power Stirling engines, a radiator is generally required to transfer the heat from the engine to the ambient air. Stirling marine engines have the advantage of using the ambient temperature water. Placing the cooling radiator section in seawater rather than ambient air allows for the radiator to be smaller. The engine's cooling water may be used directly or indirectly for heating and cooling purposes of the ship. The Stirling engine has potential for surface-ship propulsion, as the engine's larger physical size is less of a concern.

SCREWS

Marine propellers are also known as "screws". There are many variations of marine screw systems, including twin, contra-rotating, controllable-pitch, and nozzle-style screws. While smaller vessels tend to have a single screw, even very large ships such as tankers, container ships and bulk carriers may have single screws for reasons of fuel efficiency. Other vessels may have twin, triple or quadruple screws. Power is transmitted from the engine to the screw by way of a propeller shaft, which may or may not be connected to a gearbox.

Paddle wheels

Left: original paddle wheel from a paddle steamer.

Right: detail of a paddle steamer.

The paddle wheel is a large wheel, generally built of a steel framework, upon the outer edge of which are fitted numerous paddle blades (called *floats* or *buckets*). The bottom quarter or so of the wheel travels underwater. Rotation of the paddle wheel produces thrust, forward or backward as required. More advanced paddle wheel designs have featured *feathering* methods that keep each paddle blade oriented closer to vertical while it is in the water; this increases efficiency. The upper part of a paddle wheel is normally enclosed in a paddlebox to minimise splashing.

Paddle wheels have been superseded by screws, which are a much more efficient form of propulsion. Nevertheless, paddle wheels have two advantages over screws, making them suitable for vessels in shallow rivers and constrained waters: first, they are less likely to be clogged by obstacles and debris; and secondly, when contra-rotating, they allow the vessel to spin around its own vertical axis. Some vessels had a single screw in addition to two paddle wheels, to gain the advantages of both types of propulsion.

SAILING

The purpose of sails is to use wind energy to propel the vessel, sled, board, vehicle or rotor.

WATER CATERPILLAR

An early uncommon means of boat propulsion was the water caterpillar. This moved a series of paddles on chains along the bottom of the boat to propel it over the water and preceded the development of tracked vehicles. The first water caterpillar was developed by Desblancs in 1782 and propelled by a steam engine. In the United States the first water caterpillar was patented in 1839 by William Leavenworth of New York.

BUOYANCY

Underwater gliders convert buoyancy to thrust, using wings, or more recently hull shape (SeaExplorer Glider). Buoyancy is made alternatively negative and positive, generating tooth-saw profiles.

FUNCTION OF DIESEL ENGINE - PREPARATIONS FOR STANDBY, STARTING,REVERSING AND RUNNING AT FULL SPEED

MARINE DIESEL ENGINE OPERATION

The diesel engine is a type of internal combustion engine which ignites the fuel by injecting it into hot, high-pressure air in a combustion chamber. In common with all internal combustion engines the diesel engine operates with a fixed sequence of events, which may be achieved either in four strokes or two, a stroke being the travel of the piston between its extreme points. Each stroke is accomplished in half a revolution of the crankshaft.

Main Engine Power and Vibration

The normal service power of the Main Engine must be maintained as instructed in the vessel's commissioning letter unless otherwise directed by the Company, except under emergency conditions involving safety of life or safety of the ship.

Should the vessel's normal service power require to be altered the fact, together with the reason for the alteration, is to be reported to the Company and noted in the Engine Log Book. The Chief Engineer is to take instructions from the Master for the specific voyage requirements, always keeping within safe operating parameters.

Vibration can cause severe damage to machinery, bearings, pipes, fittings, instrumentation and structure. To minimise this damage the main machinery must be regulated at all times to avoid speeds at which excessive vibration may be experienced.

Besides the barred-speed ranges prescribed by the engine designers, operation at certain speeds where the combination of draught, trim and weather leads to severe vibration, is also to be avoided.

Particular attention must be paid to the balancing of cylinder loads in Diesel engines and to the tightness of holding down bolts on all reciprocating machinery. Full use must be made of all condition monitoring equipment supplied to detect and measure vibration, and any significant increase in vibration levels that cannot be accounted for must be reported to the management ashore.

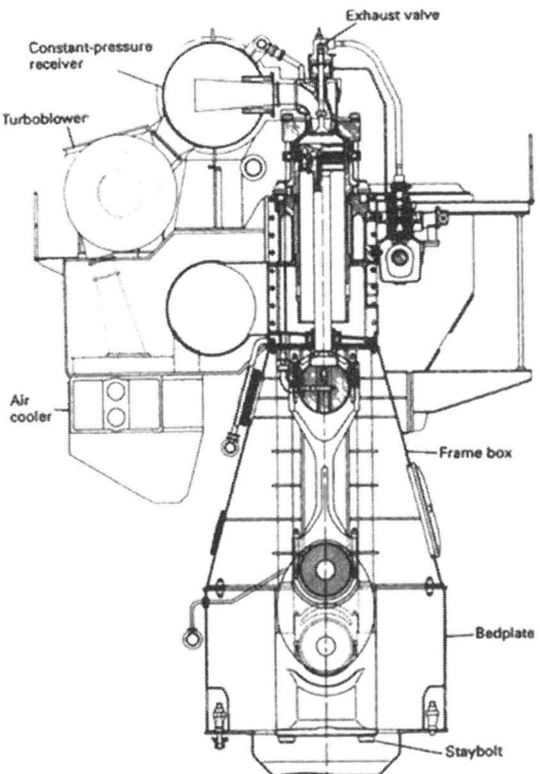

Fig. MAN B&W L70MC engine

WARMING THROUGH

Main engines are to be warmed through gradually following a stay in port or other occasion when they have been shut down. The jacket water circulation temperature is to be raised over a period of time to as near to the operating temperature as possible. The period of time is dependent on the jacket water temperature prior to the commencement of circulation, the heating medium and size of main engine etc. As a general rule circulation should commence not less than 12 hours before the estimated time of departure. Other circulating systems are to be put on line during this period i.e.

- Lubricating oil systems.
- Fuel circulating systems.
- Steam tracing systems as is appropriate to the type of engine.

PRECAUTIONS PRIOR TO STANDBY DEPARTURE

All circulating systems are to be as near as is possible normal operating parameters according to the manufacturer's instructions for the engine type. On vessels where the engine or engines are directly coupled to the propeller or propellers, the Deck Officer of the watch or Duty Deck Officer is to be contacted and permission to turn the engine requested. Once the Deck Officer has confirmed that this can safely be carried out, then and only then can the engine be turned.

All the indicator cocks are to be opened. The engine is then to be turned a minimum of one revolution with the turning gear during which time the indicator cocks are to be sighted for any evidence of oil or water discharge. If this test is satisfactory, then the indicator cocks are to be shut and the turning gear disengaged. It is of the utmost importance that the disengagement of the turning gear is physically checked. No reliance is to be placed on the indicator light in the control room, interlocks etc. It is the Chief Engineer's responsibility to ensure that this physical check is carried out.

On vessels where the engine or engines are directly coupled to the propeller or propellers, the Deck Officer of the watch or Duty Deck Officer is to be contacted and permission to turn the engine on air requested. Once the Deck Officer has confirmed that this can safely be done, then and only then can the engine be turned on air. All indicator cocks are to be opened. In conjunction with the Bridge and as applicable, the engine is to be "kicked" ahead and astern on starting air. The indicator cocks are then to be closed.

Following the satisfactory turning of the engine on air, the Deck Officer of the watch is to be contacted and permission to turn the engine on fuel requested. Once the Deck Officer has confirmed that this can safely be done then, and only then, can the engine be turned on fuel. In conjunction with the Bridge the engine is to be turned dead slow ahead and astern on fuel.

OPERATION ON HEAVY FUEL OIL

Main engines designed to manoeuvre on heavy fuel oil are to be operated according to the manufacturer's instructions. In the event of problems during manoeuvring on engines using heavy oil there must be no hesitation in changing over to diesel oil irrespective of whether the engines are being operated using bridge control, or using engine room control. It is the Chief Engineer's responsibility to inform the Master of the particular engine type's maximum period that it can safely remain in the stopped position. He is also to inform the Master of the procedures which will have to be carried out if the particular engine type's maximum period at standstill during manoeuvring is exceeded.

Preparations for standby

1. Before a large diesel is started it must be warmed through by circulating hot water through the jackets, etc. This will enable the various engine parts to expand in relation to one another.
2. The various supply tanks, filters, valves and drains are all to be checked.

3. The lubricating oil pumps and circulating water pumps are started and all the visible returns should be observed.
4. All control equipment and alarms should be examined for correct operation.
5. The indicator cocks are opened, the turning gear engaged and the engine turned through several complete revolutions. In this way any water which may have collected in the cylinders will be forced out.
6. The fuel oil system is checked and circulated with hot oil.
7. Auxiliary scavenge blowers, if manually operated, should be started.
8. The turning gear is removed and if possible the engine should be turned over on air before closing the indicator cocks.
9. The engine is now available for standby.

The length of time involved in these preparations will depend upon the size of the engine.

Engine starting

1. The direction handle is positioned ahead or astern. This handle may be built into the telegraph reply lever. The camshaft is thus positioned relative to the crankshaft to operate the various cams for fuel injection, valve operation, etc.
2. The manoeuvring handle is moved to 'start'. This will admit compressed air into the cylinders in the correct sequence to turn the engine in the desired direction.A separate air start button may be used.
3. When the engine reaches its firing speed the manoeuvring handle is moved to the running position. Fuel is admitted and the combustion process will accelerate the engine and starting air admission will cease.

Engine reversing

When running at manoeuvring speeds:

1. Where manually operated auxiliary blowers are Fitted they should be started.
2. The fuel supply is shut off and the engine will quickly slow down,
3. The direction handle is positioned astern.
4. Compressed air is admitted to the engine to turn it in the astern direction.
5. When turning astern under the action of compressed air, fuel will be admitted. The combustion process will take over and air admission cease.

When running at full speed:

1. The auxiliary blowers, where manually operated, should be started.

2. Fuel is shut off from the engine.
3. Blasts of compressed air may be used to slow the engine down.
4. When the engine is stopped the direction handle is positioned astern.
5. Compressed air is admitted to turn the engine astern and fuel is admitted to accelerate the engine. The compressed air supply will then cease.

FUNCTION OF PISTON IN A MARINE DIESEL ENGINE

PISTON FORMING THE LOWER PART OF THE COMBUSTION CHAMBER

The diesel engine is a type of internal combustion engine which ignites the fuel by injecting it into hot, high-pressure air in a combustion chamber. In common with all internal combustion engines the diesel engine operates with a fixed sequence of events, which may be achieved either in four strokes or two, a stroke being the travel of the piston between its extreme points. Each stroke is accomplished in half a revolution of the crankshaft.

Piston forms the lower part of the combustion chamber. It seals the cylinder and transmits the gas pressure to the connecting rod. The piston absorbs heat of combustion and this heat must be conducted away if the metal temperature is to kept in safe limits. The Piston comprises of two pieces; the crown and the skirt.

The crown is subject to the high temperatures in the combustion space and the surface is liable to be eroded/burnt away. For this reason the material from which the crown is made must be able to maintain its strength and resist corrosion at high temperatures.

Fig. Handling piston

Steel, alloyed with chromium and molybdenum is used, and some pistons have a special alloy welded onto the hottest part of the crown to try and reduce the erosion caused by the burning fuel. The crown also carries the 4 or 5 piston ring grooves which may be chrome plated.

The cast iron skirt acts as a guide within the cylinder liner. It is only a short skirt on engines with an exhaust valve (known as uniflow scavenged engines), as unlike a trunk piston engine, no side thrust is transmitted to the liner (that's the job of the crosshead guides).

The stresses to which a piston is subjected to are as follows:

- (mechanical & thermal stresses)
- Compressive and tensile stress caused by bending action due to gas pressure
- Inertia effect - movement up and downwards
- Thermal stresses - rapid temperature change

The crown of a piston is subjected to a very high gas pressure which will subject the top surface of the crown to compressive loading and the lower surface of crown will be under tensile loading. The piston crown will be like a uniformly loaded beam

As the piston moving upward, towards the end of its stroke its velocity will be reducing.The inertia effect will tend to cause the piston to bow upwards, so that the surface of the crown along with sides will be under tensile loading and lower surface of the crown will be under compressive loading When the piston is retarded on its approach downwards to BDC, the inertia effect will be reversed

The thermal stresses set up in a piston are caused by the different temperature across a section. The free expansion of the hot side is restricted by the cooler surface of the piston. Maximum safe temperature at the three most critical zones for alluminium-alloy piston are Crown 370 degree to 400 degree C. Top ring groove and gudgeon pin bosses 200 degree to 220 degree C. Mechanical and thermal stresses should be considered to gather as they tend to be complimentary to each other. Top and sometimes 2nd ring grooves is tapered up to 2 times normal axial clearance thus:

- Clearance allows for carbon deposit

Clearance avoids piston crown edge touching liner due to thermal stress

If the crown temperature exceeds 400 degreeC, failure will probably occur from cracking. If the top ring groove temperature exceeds 220 degreeC, for any length of time, trouble may be expected from:

- Stuck piston rings
- Formation of carbon at the bottom of the ring groove, causing the ring to be packed out

Cracking of crown – due to thermal and mechanical stresses. Cracking through piston wall especially in way of top ring groove – due to fluctuating gas load, excessive thermal stresses. Crack starts from inside wall. Cracking may take place due to the following reasons apart from the reasons mention above:

- Unsuitable material for the rating of the engine or inadequate machining
- Excessive scaling on the cooling side, cavitaion erosion
- High coolant temperature
- Local impingement
- Poor atomization, high penetration of fuel
- High water content in fuel

CONSTRUCTION OF 4 STROKE PISTON IN A DIESEL ENGINE

piston cooling process

Piston forms the lower part of the combustion chamber of marine diesel engine. Cast iron is the most common material used for piston. Cracking of cast iron piston much reduced by the use of iron castings of pearlitic structure, with less tendency to growth; as well as oil cooling of the larger sizes of piston.

Pistons for medium speed trunk piston engines which burn residual fuel are composite pistons; i.e the crown and the skirt are made of different materials. Piston crowns attain a running temperature of about 450 degree C and in this zone there is a need for high strength and minimum distortion in order to maintain resistance to gas loads and maintain the attitude to the rings in relation to the liner. The heat flow path from the crown must be uniform otherwise thermal distortion will cause a non-circular piston resulting in reduced running clearance or even possible contact with the liner wall. In addition to this thermal stress they are also subject to compressive stress from combustion and compression loads, as well as inertial loads.

Materials such as pearlitic, flake and spheroidal cast iron, alloy cast irons containing Nickel and chromium, and aluminium alloys may be used. The determining factor is the design criteria for the engine.

For a modern slow speed engine steel forging or castings of nickel-chrome steel or molybdenum steel are common. The weight of the material is not normally a governing factor in this type of engine although resistance to thermal stress and distortion is. Efficient cooling is a required to ensure the piston retains sufficient strength to prevent distortion.

For medium and high speed engines the weight of the material becomes important to reduce the stresses on the rotating parts. The high thermal conductivity of aluminium alloys allied to its low weight makes this an ideal material. To keep thermal stresses to a reasonable level cooling pipes may be cast into the crown, although this may be omitted on smaller engines.Where cooling is omitted, the crown is made thicker both for strength and to aid in the heat removal from the outer surface.

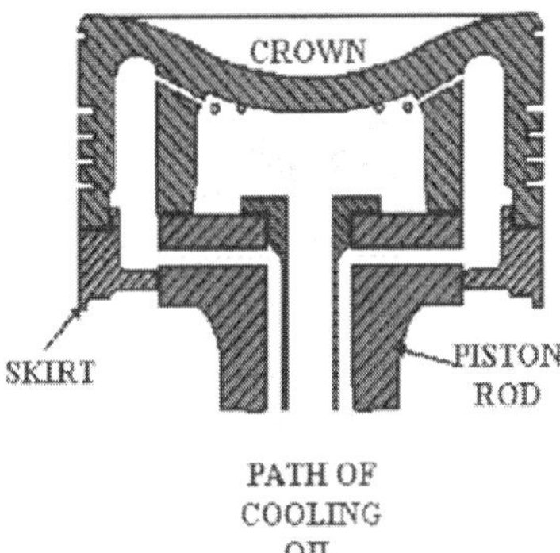

Fig. Path of piston cooling

Hard landings are inserted into the ring groves to keep wear rated down.Composite pistons may be used consisting of an cast alloy steel crown with an aluminium-alloy or cast iron body.

After casting or forging the component is formed of different material thicknesses. The thinner parts will cool more quickly thereby setting up internal stresses. Annealing removes or reduces these stresse as well as refining the grain structure.

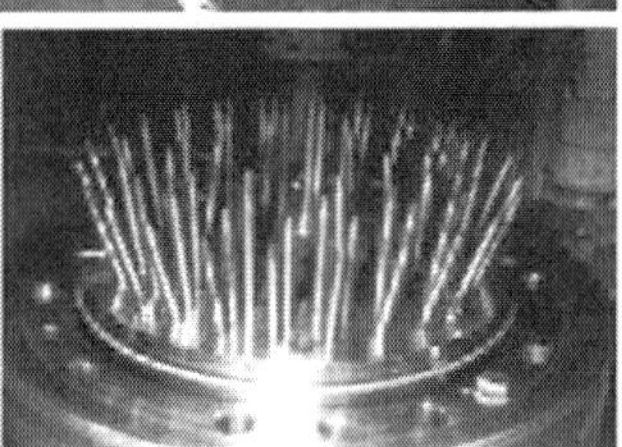

This Photo Shows the Nozzle Plate and Nozzles

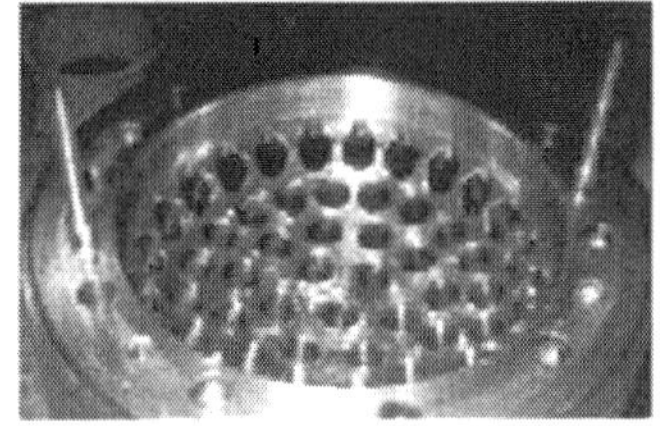

This Photo Shows the Underside of the Piston Crown

Fig. Piston crown

The crown is a heat resisting steel forging which may be alloyed with chromium, molybdenum and nickel to maintain strength at high temperatures and resist corrosion. It is dished to form a combustion chamber with cutouts to allow for the valves opening. The topland (the space between the top ring and the top of the piston) may be tapered to allow for expansion being greater where the piston is hottest.

The skirt can either be a nodular cast iron or forged or cast silicon aluminium alloy. Aluminium has the advantage of being light, with low inertia, reducing bearing loading. However because aluminium has a higher coefficient of expansion than steel, increased clearances must be allowed for during manufacture. This means that the piston skirt clearance in the liner is greater than that for cast iron when running at low loads. The skirt transmits the side thrust, caused by the varying angularity of the con rod, to the liner. Too big a clearance will cause the piston to tilt.

The piston pin for the con rod small end bearing is located in the piston skirt. The piston pin floats in the piston skirt and is located in place by circlips. Depending on the material used for the skirt (esp. cast aluminium), a bushing may be used for the pin.

The piston rings may be located in the crown or in both crown and skirt. Normally, the rings are chrome plated or plasma coated to resist wear. Because the liner is splash lubricated, an oil scraper (oil control) ring is fitted to the piston skirt. The piston is oil cooled. This is achieved by various means; The simplest is for a jet of oil to be directed upwards from a hole in the top of the con rod onto the underside of the crown

PISTON COOLING METHODS - ADVANTAGE AND DISADVANTAGES OF WATER COOLED AND OIL COOLED PISTONS

COOLANT USED FOR COOLING A PISTON OF MARINE DIESEL ENGINE

The coolant used for removing and conveying the heat from a piston may be either fresh water, distilled water or lubricating oil. Water has the ability to remove more heat than lubricating oil (specific heat of water approximately 4 and lubricating oil 2 and temperature difference 14oC for water and 10oC for lube oil) Modern engines have oil cooled pistons. The piston rod is utilised to carry the oil to and from the piston. The rod is hollow, and has a tube running up its centre. This gives an annular space which, with the central bore, allows a supply and return.

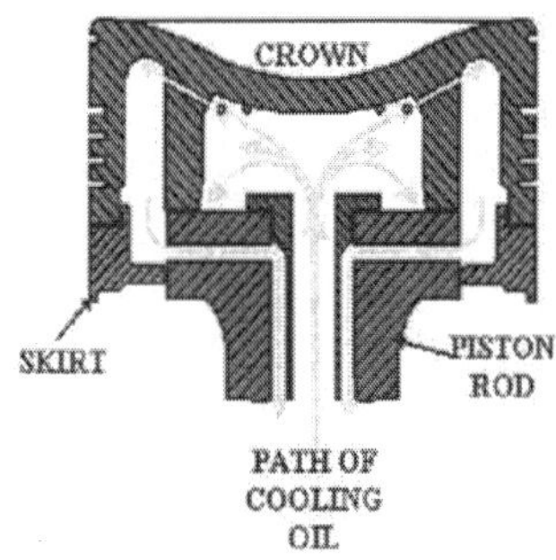

Fig. Path of piston oil cooling

Fresh water and distilled water piston cooling advantages:

- The main advantages is the ability of water to absorb large amounts of heat Relatively easy to obtain.
- Easily counter check from drain tank any leakage on piston sealing.

Disadvantages:

- The piston cooling water conveyance piston and attendant gear must be kept out of the crankcase as far as possible, due to the danger of contamination of the crankcase lubricating oil by water leakage. In other word, the jacket cooling water and piston cooling system should have separate which having their own pumps, coolers, piping etc.
- The piston cooling space should be drained of water after the engine is shut down for an extend period. A drain tank is necessary for the same purpose cascade type filter is often incorporated for separation of oil and water.
- There is risk of scaling and corrosion if water is not suitable treated and maintained.

Lubricating oil piston cooling Advantages:

- The piston cooling pump is combined with the lubricating oil pump and piston cooling oil cooler thus make overall simplicity in the system
- Internal stress within the material of the piston is generally less in oil cooled piston than in water-cooled piston. Good design in water-cooled piston to improve its condition of working
- No risk of oil contamination even when piston cooling oil conveyance piping is fitted inside the crankcase
- Simpler arrangement for cooling-oil conveyance piping with less risk of 'hammering' in piping and bubble impingement attack

Disadvantages:

- Larger power requirements for pumping cooling oil
- Larger amount of lubricating oil required to give some cooling effect

FLOW PATTERN OF THE COOLANT

The flow is such that piston cooling oil or water enters at the lowest part of the cooling space and leaves from the uppermost part to give even cooling effect without causing distortion due to unequal expansion.

The piston will always full of coolant and the underside of piston crown is always in contact with it It is important in slow speed engine, as when the piston is running at dead slow speed the coolant in the piston is not 'shaken up' the way the engine in full speed.

If the cooling took place in the opposite direction, it would be possible at very slow speed for the coolant to drain from the piston and lose contact with the crown. The piston could become overheated.

Some water-cooled piston have the outlet for the water at approximately half the cooling space height. When running slow, the piston is half full of water and piston movement agitates this water in the piston – the water gets splashed on the underside of crown, piston wall

When the engine stopped a jet action from the piston cooling pipe nozzle directs cooling water onto the piston crown, thus removing residual heat and catering for an emergency stop at full speed. The splash method of cooling is called "cocktail shaker cooling"

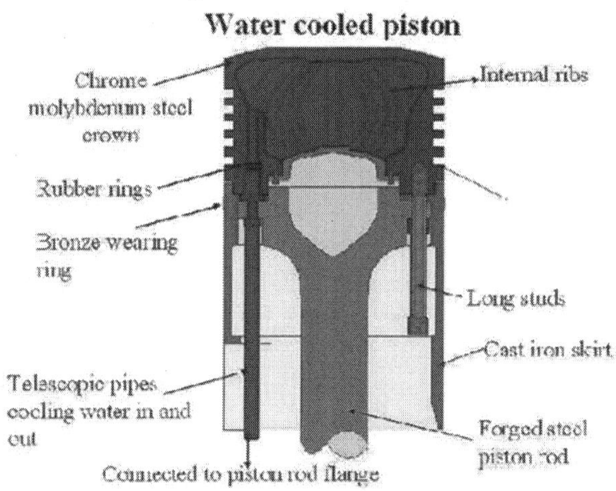

Fig. Water cooled piston

REQUIREMENT OF PISTON RINGS IN A MARINE DIESEL ENGINE

SEAL THE CYLINDER, DISTRIBUTE AND CONTROL LUBRICATING OIL ON THE CYLINDER WALL BY PISTON RING

Piston rings are particularly vital to marine diesel engine operation in that they must effectively perform three functions: seal the cylinder, distribute and control lubricating oil on the cylinder wall, and transfer heat from the piston to the cylinder wall. All rings on a piston perform the latter function, but two general types of rings—compression and oil—are required to perform the first two functions. The number of rings and their location will also vary considerably with the type and size of the piston.

Rings must have sufficient spring so that they will provide an initial seal with the liner. As pressure builds up gas acting on the back face of the ring increase the sealing effect. The spring must be retained under normal operating temperatures. They must not crack under high temperature and pressure ranges. Rings are generally of spherical graphite cast iron because of the strength and limited self lubricating properties. With modern long stroke engines the rings do considerably more rubbing than equivalent sections of the liner and so the rubbing faces are usually made slightly harder. This is achieved by a case hardening process (usually Nitriding) some rings are contoured on the rubbing face in order to promote faster running in.

Copper or carbon coatings are sometimes provided for the same purpose. When running in cylinder l.o. is increased to provide an additional flow to carry away metallic particles and a straight mineral oil without antiwear properties is used. The ring axial depth must be sufficient to provide a good seal against the liner but it must not be so great so that an oil wedge does not form. The ring actually distorts in the groove to form the wedge but if they are too deep they cannot do so. Thin rings will distort easily and scrape the oil from the surface. Radial depth must be sufficient to allow adequate support for the ring in the groove when the ring is on max. normal wear for its self and the liner.

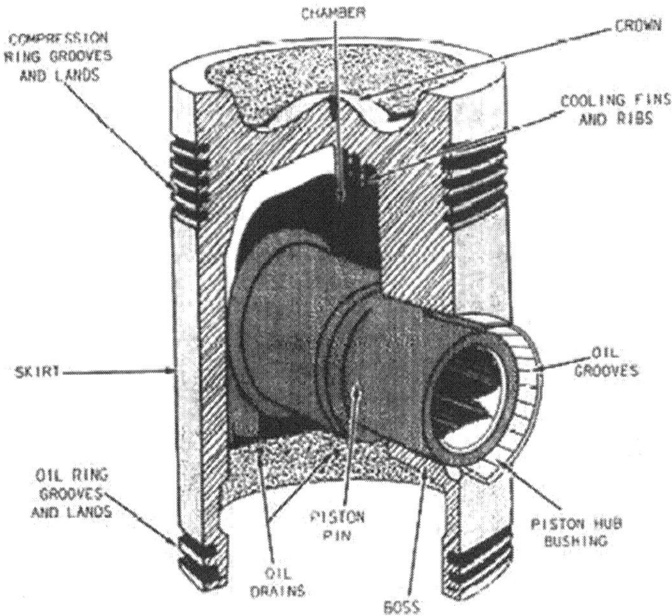

Fig. piston-configuration

Rings must be free in their grooves and the correct clearance is required. Excessive clearance can allow rings to twist while insufficient clearance can cause jamming and prevent the gas pressure from acting behind the rings. Also the rings may tend to twist excessively. Radial clearance must be sufficient between groove and ring back to allow a gas cushion to build up. The butt clearance must be sufficient to allow for thermal expansion. If insufficient the rings may seize and if excessive can lead to excessive blowpast

Grooves are sometimes coated with chromium to restrict deposit build up. For reconditioning the bottom face of the groove is generally provided with a replaceable steel wear ring. As the rings maintain the gas seal there is a desire to position the top or firing ring as close to the piston crown as possible. However,since the crown is highly stressed, thermally, this results in distortion of that zone. There is thus a desire to position the ring a long distance away from the crown. A compromise position is decided upon in each engine design.

In order to minimise wear, a film of lubricating oil must be maintained between the moving parts i.e. the rings and liner, and rings and groove. Also the lubricating oil must spread over the liner surface by the rings, this helps to combat acidic products of combustion.

Skirts fitted to pistons on some designs perform the function of sealing the exhaust ports at T.D.C. these extended skirts have bronze rubbing rings inset to provide a bearing surface during the running in period.

FUNCTION OF PISTON RING IN A MARINE DIESEL ENGINE

HOW PISTON RING WORKS AS EFFECTIVE SEALING BETWEEN THE PISTON AND LINERS

The diesel engine is a type of internal combustion engine which ignites the fuel by injecting it into hot, high-pressure air in a combustion chamber.

In common with all internal combustion engines the diesel engine operates with a fixed sequence of events, which may be achieved either in four strokes or two, a stroke being the travel of the piston between its extreme points. Each stroke is accomplished in half a revolution of the crankshaft.

The efficiency of the engine depends upon the effective sealing between the piston and liners. Leakage will reduced compression pressure and power will lost. Piston rings seal the gas space by expanding outwards due to the gas pressure acting behind them. They also spread the lubricating oil up and down the cylinder liner and transfer heat to the liner walls Three to six power or compression rings are fitted to the piston, the number depending on weather the engine operates on the 2-stroke or 4-stroke cycle.

The top ring position is controlled by the operating temperature and exhaust or scavenge port trimming. In 2-stroke cycle engines the port trimming usually demands a higher ring position than used on 4- stroke cycle engine The main function of piston rings are;

- To provide a gas seal in the case of compression rings and to prevent excessive built up of pressure in the crankcase which will results inexcessive lube oil consumption in a wet sump engine and the risk of explosion in the case of a dry sump unit
- To act as a scrapper or oil distribution around liner wall which could result if lubricant is allowed to leak past the compression rings and be burnt with fuel oil

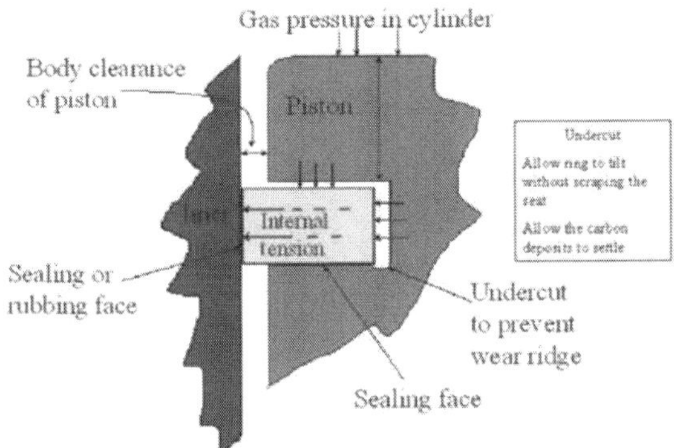

Fig. Piston ring sealing face

To ensure adequate lubrication the operating temperature should not exceed 200oC and this requires the rings to be fitted as low down the piston as possible. Lubrication of the rings is essential as it is needed to reduce friction and wear to augment gas sealing Ring will normally rotate on piston due to the variation in gas pressure. A typical rate of movement is 1 – 3 mm per cycle

The diameter of the piston rings, when free, is slightly larger than the cylinder bore. Consequently, when the ring is squeezed into the liner, it presses against the cylinder wall and tend to seal it.

This initial sealing action is greatly improved by the pressure either compression or combustion gasses against the top of the ring forces the ring down on to the lower side of the piston grooves. This leaves a clearance at the top which permits gas pressure to travel behind the back of the ring and press it against the cylinder liner wall making a firmer contact, thus improving the seal

There are three clearances which are important;-

- Side or axial clearances
- Butt or Gap clearances
- Back clearances

HOW A TWO-STROKE CYCLE DIESEL ENGINE WORKS

MARINE DIESEL ENGINE GUIDELINE

The two-stroke cycle is completed in two strokes of the piston or one revolution of the crankshaft. In order to operate this cycle where each event is accomplished in a very short time, the engine requires a number of special arrangements. First, the fresh air must be forced in under pressure. The incoming air is used to clean out or scavenge the exhaust gases and then to fill or charge the space with fresh air. Instead of valves holes, known as 'ports', are used which are opened and closed by the sides of the piston as it moves.

A cross-section of a two-stroke cycle engine is shown in Figure. The piston is solidly connected to a piston rod which is attached to a crosshead bearing at the other end. The top end of the connecting rod is also joined to the crosshead bearing.

Ports are arranged in the cylinder liner for air inlet and a valve in the cylinder head enables the release of exhaust gases.

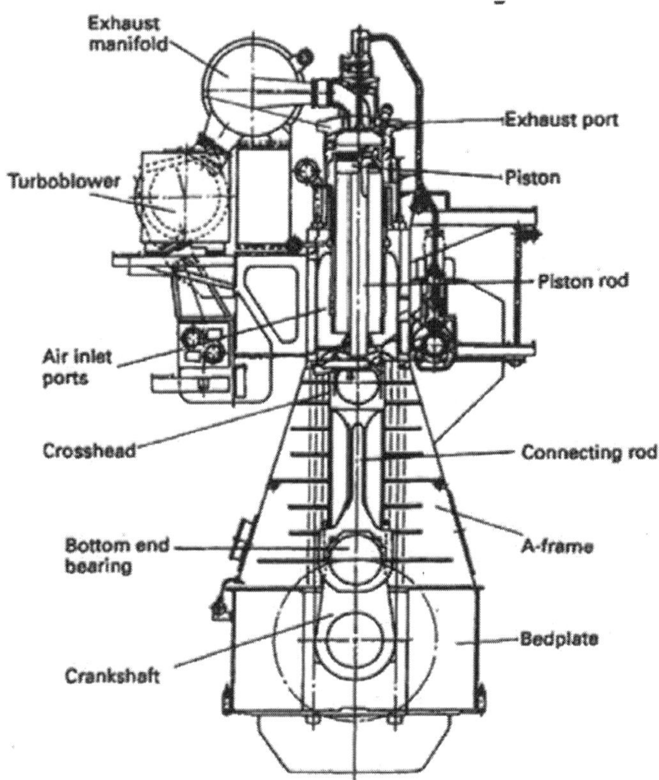

The incoming air is pressurised by a turbo-blower which is driven by the outgoing exhaust gases. The crankshaft is supported within the engine bedplate by the main bearings. A-frames are mounted on the bedplate and house guides in which the crosshead travels up and down. The entablature is mounted above the frames and is made up of the cylinders, cylinder heads and the scavenge trunking.

COMPARISON BETWEEN TWO STROKE CYCLE DIESEL ENGINE AND A FOUR STROKE ENGINE

MARINE DIESEL ENGINE GUIDELINE

The main difference between the two cycles is the power developed. The two-stroke cycle engine, with one working or power stroke every revolution, will, theoretically, develop twice the power of a four-stroke engine of the same swept volume. Inefficient scavenging however and other losses, reduce the power advantage. For a particular engine power the two-stroke engine will be considerably lighter—an important consideration for ships. Nor does the two-stroke engine require the complicated valve operating mechanism of the four-stroke. The four-stroke engine however can operate efficiently at high speeds which offsets its power disadvantage; it also consumes less lubricating oil.

Each type of engine has its applications which on board ship have resulted in the slow speed (i.e. 80— 100 rev/min) main propulsion diesel operating on the two-stroke cycle. At this low speed the engine requires no reduction gearbox between it and the propeller.

The four-stroke engine (usually rotating at medium speed, between 250 and 750 rev/ min) is used for auxiliaries such as alternators and sometimes for main propulsion with a gearbox to provide a propeller speed of between 80 and 100 rev/min.

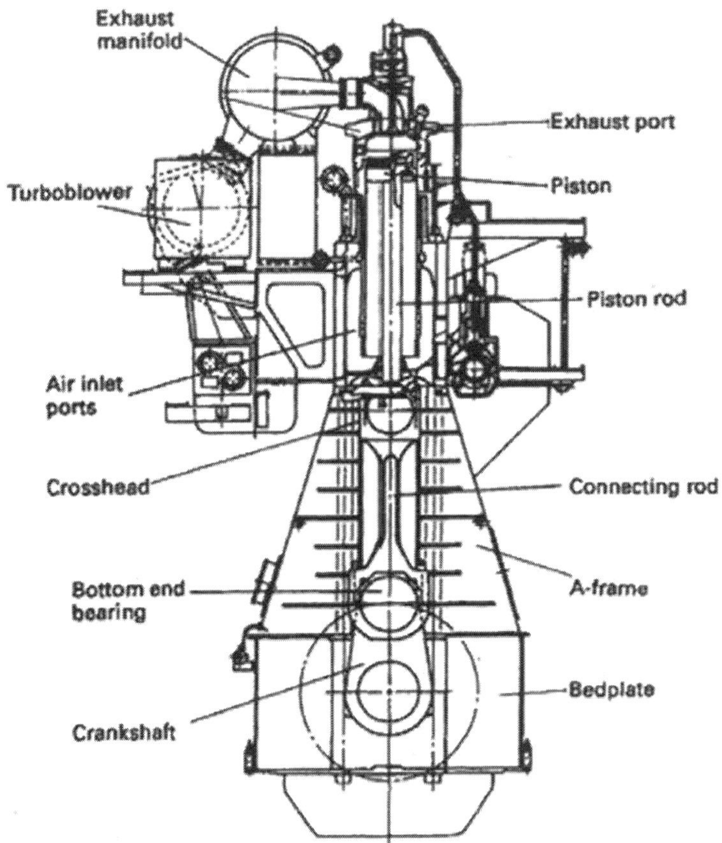

Fig. Two stroke engine

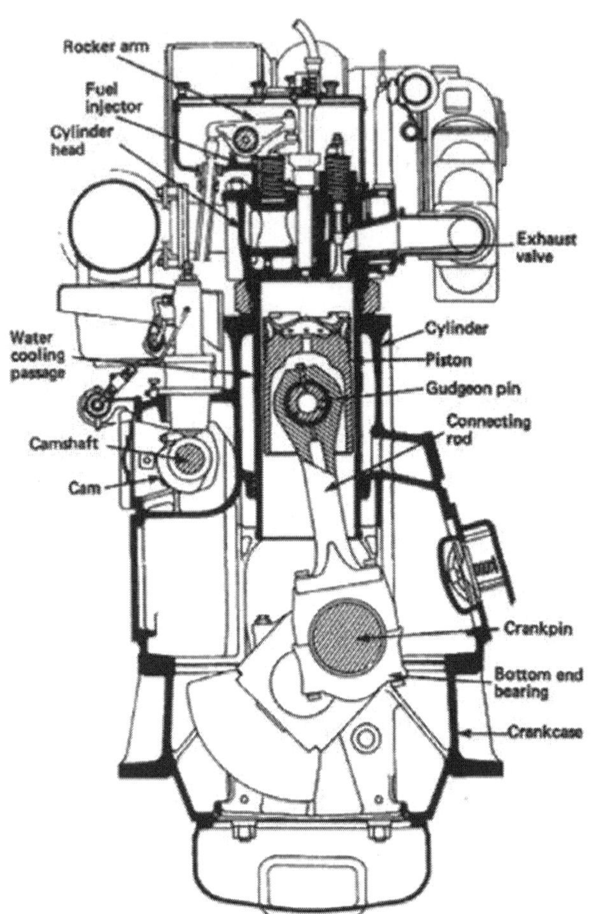

Fig. Four stroke engine

POWER MEASUREMENT FOR MARINE DIESEL ENGINE - THE ENGINE INDICATOR

MARINE DIESEL ENGINE POWER MEASUREMENT GUIDELINE

There are two possible measurements of engine power: the indicated power and the shaft power. The indicated power is the power developed within the engine cylinder and can be measured by an engine indicator. The shaft power is the power available at the output shaft of the engine and can be measured using a torsionmeter or with a brake.

An engine indicator is shown in Figure below. It is made up of a small piston of known size which operates in a cylinder against a specially calibrated spring. A magnifying linkage transfers the piston movement to a drum on which is mounted a piece of paper or card. The drum oscillates (moves backwards and forwards) under the pull of the cord. The cord is moved by a reciprocating (up and down) mechanism which is proportional to the engine piston movement in the cylinder.

The stylus draws out an indicator diagram which represents the gas pressure on the engine piston at different points of the stroke, and the area of the indicator diagram produced represents the power developed in the particular cylinder. The cylinder power can be measured if the scaling factors, spring calibration and some basic engine details are known. The cylinder power values are compared, and for balanced loading should all be the same. Adjustments may then be made to the fuel supply in order to balance the cylinder loads.

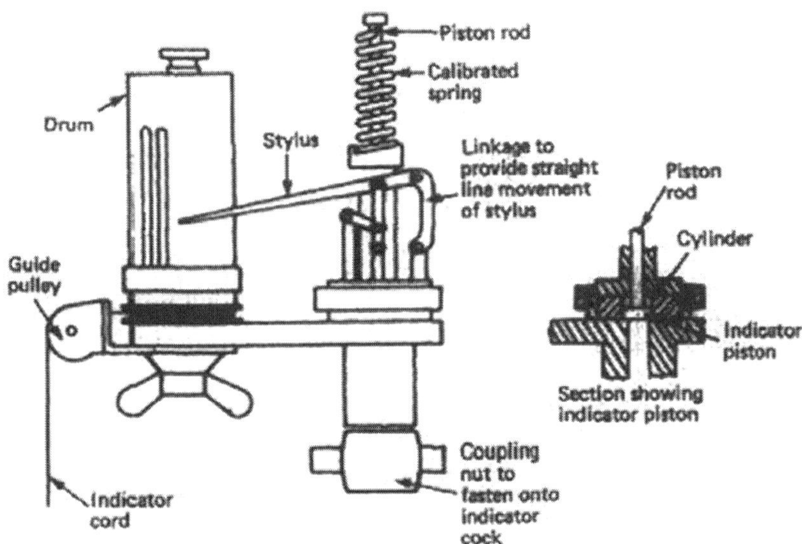

TORSIONMETER

If the torque transmitted by a shaft is known, together with the angular velocity, then the power can be measured, i.e. shaft power = torque x angular velocity. The torque on a shaft can be found by measuring the shear stress or angle of twist with a torsionmeter.

THE MEASUREMENT OF PRESSURE BY MANOMETER OR A BAROMETER

PRESSURE MEASUREMENT BY SHIPBOARD MACHINERY

Shipboard machinery must operate within certain desired parameters. Instrumentation enables the parameters—pressure, temperature, and so on— to be measured or displayed against a scale. A means of control is also required in order to change or alter the displayed readings to meet particular requirements. Control must be manual, the opening or closing of a valve, or automatic, where a change in the system parameter results in actions which return the value to that desired without human involvement.

PRESSURE MEASUREMENT

The measurement of pressure may take place from one of two possible datums, depending upon the type of instrument used. Absolute pressure is a total measurement using zero pressure as datum. Gauge pressure is a measurement above the atmospheric pressure which is used as a datum. To express gauge pressure as an absolute value it is therefore necessary to add the atmospheric pressure.

MANOMETER

A U-tube manometer is shown in Figure. One end is connected to the pressure source; the other is open to atmosphere. The liquid in the tube may be water or mercury and it will be positioned as shown.

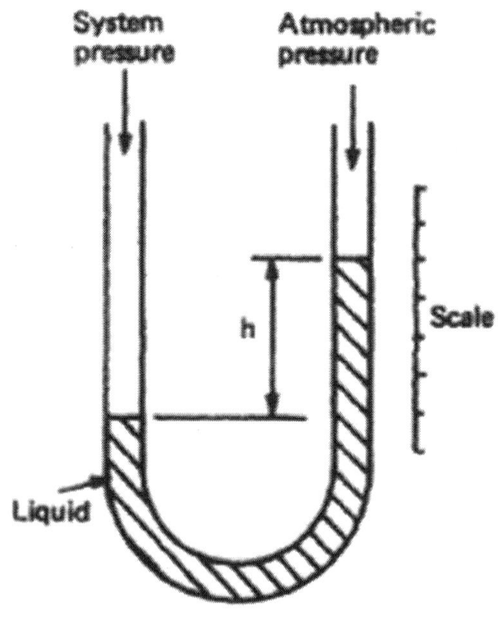

Fig. U-tube manometer

The excess of pressure above atmospheric wil be shown as the difference in liquid levels; this instrument therefore measures gauge pressure. It is usually used for low value pressure readings such as air pressures. Where two different system pressures are applied, this instrument will measure differential pressure.

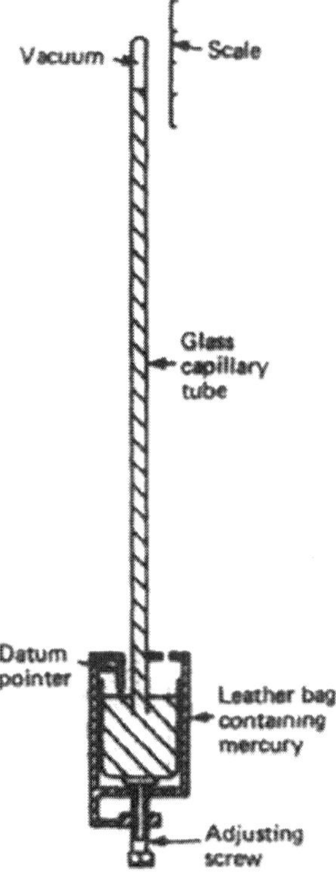

Fig. Mercury barometer

BAROMETER

The mercury barometer is a straight tube type of manometer. A glass capillary tube is sealed at one end^ filled with mercury and then inverled in a small bath of mercury. Almost vacuum conditions;exist above the column of mercury, which is supported by atmospheric pressure acting on the mercury in the container. An absolute reading of atmospheric pressure is thus given. The aneroid barometer uses an evacuated corrugated cylinder to detect changes in atmospheric pressure. The cylinder centre tends to collapse as atmospheric pressure increases or is lifted by the spring as atmospheric pressure falls. A series of linkages transfers the movement to a pointer moving over a scale.

BOURDON TUBE

This is probably the most commonly used gauge pressure measuring instrument and is shown in Figure. It is made up of an elliptical section tube formed into a C-shape and sealed at one end. The sealed end, which is free to move, has a linkage arrangement which will move a pointer over a scale. The applied pressure acts within the tube entering through the open end, which is fixed in place.

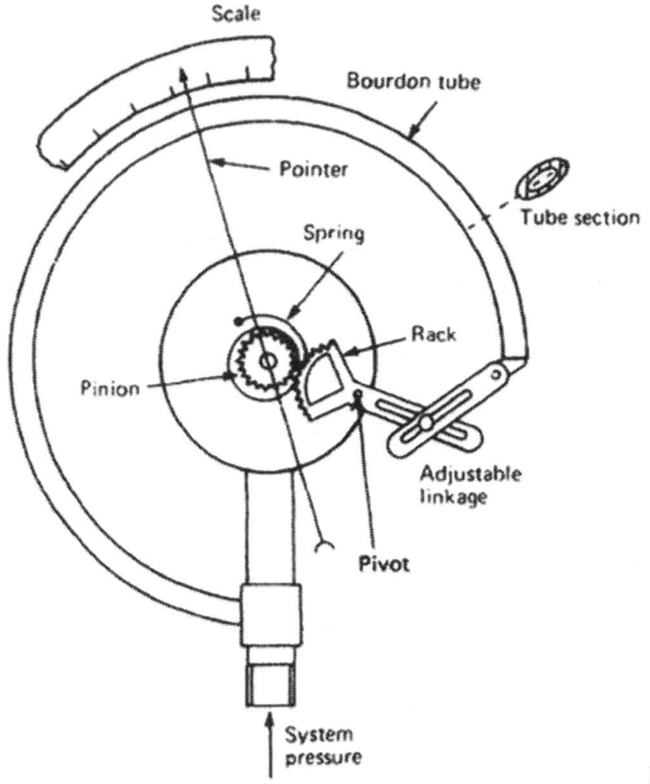

Fig. Bourdon tube pressure gauger

The pressure within the tube causes it to change in cross section and attempt to straighten out with a resultant movement of the free end, which registers as a needle movement on the scale. Other arrangements of the tube in a helical or spiral form are sometimes used, with the operating principle being the same.

While the reference or zero value is usually atmospheric, to give gauge pressure readings, this gauge can be used to read vacuum pressure values.

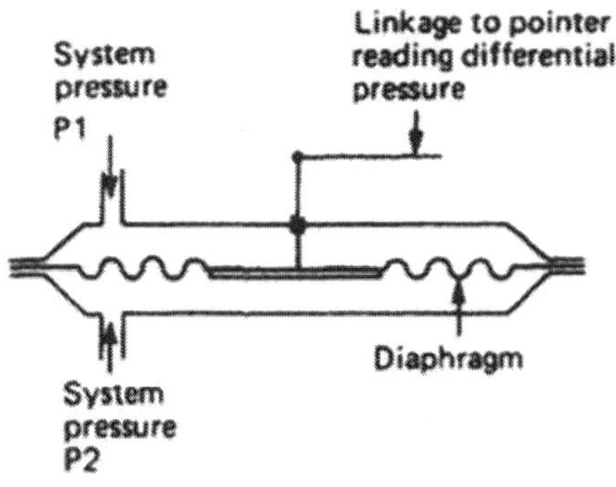

Fig. Diaphragm pressure gauge

OTHER DEVICES

Diaphragms or bellows may be used for measuring gauge or differential pressures. Typical arrangements are shown in Figure above. Movement of the diaphragm or bellows is transferred by a linkage to a needle or pointer display.

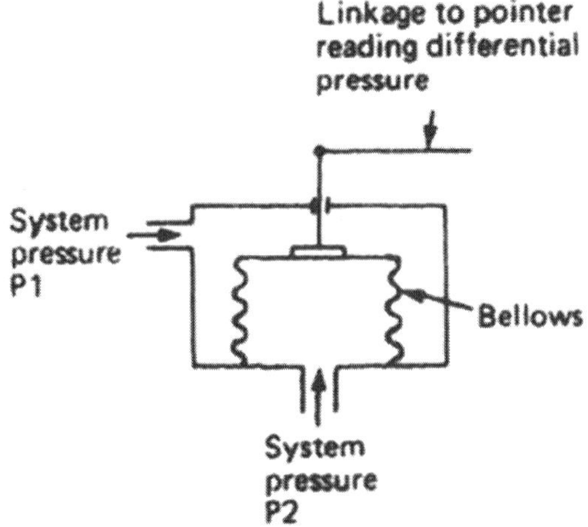

Fig. Bellows pressure gauge

The piezoelectric pressure transducer is a crystal which, under pressure, produces an electric current which varies with the pressure. This current is then provided to a unit which displays it as a pressure value.

FUNCTION OF GOVERNORS CONTROLLING SPEED OF MARINE DIESEL ENGINE

CONTROL AND SAFETY DEVICES FOR MARINE DIESEL ENGINE

The principal control device on any engine is the governor. It governs or controls the engine speed at some fixed value while power output changes to meet demand. This is achieved by the governor automatically adjusting the engine fuel pump settings to meet the desired load at the set speed.

Governors for diesel engines are usually made up of two systems: a speed sensing arrangement and a hydraulic unit which operates on the fuel pumps to change the engine power output.

MECHANICAL GOVERNOR

A flyweight assembly is used to detect engine speed. Two flyweights are fitted to a plate or ballhead which rotates about a vertical axis driven by a gear wheel. The action of centrifugal force throws the weights outwards; this lifts the vertical spindle and compresses the spring until an equilibrium situation is reached. The equilibrium position or set speed may be changed by the speed selector which alters the spring compression.

As the engine speed increases the weights move outwards and raise the spindle; a speed decrease will lower the spindle. The hydraulic unit is connected to this vertical spindle and acts as a power source to move the engine fuel controls.

A piston valve connected to the vertical spindle supplies or drains oil from the power piston which moves the fuel controls depending upon the flyweight movement. If the engine speed increases the vertical spindle rises, the piston valve rises and oil is drained from the power piston which results in a fuel control movement. This reduces fuel supply to the engine and slows it down. It is, in effect, a proportional controller.

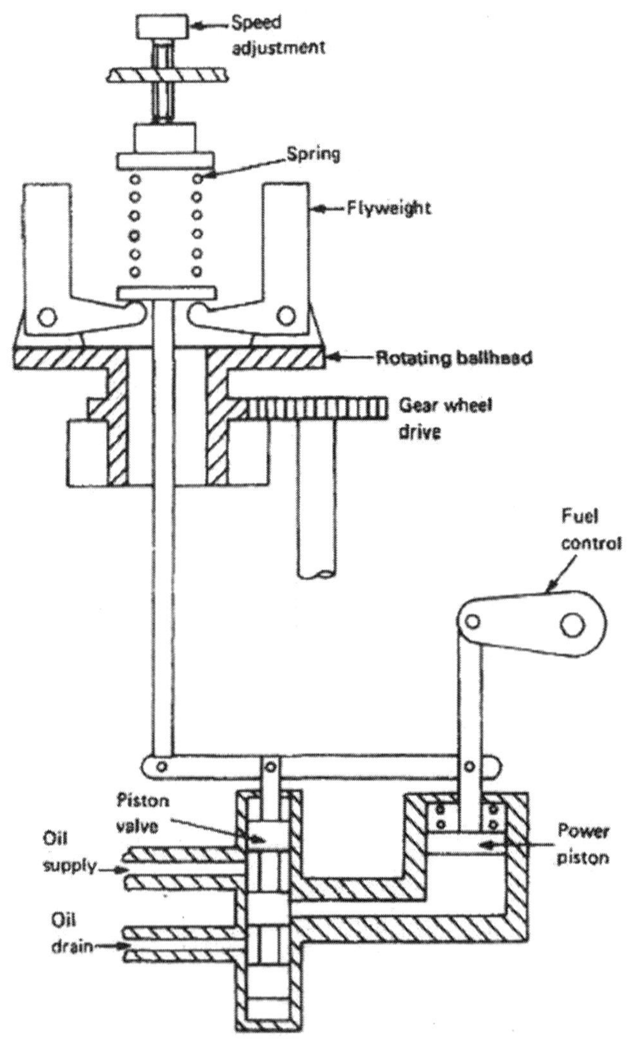

The actual arrangement of mechanical engine governors will vary considerably but most will operate as described above.

ELECTRIC GOVERNOR

The electric governor uses a combination of electrical and mechanical components in its operation. The speed sensing device is a small magnetic pick-up coil. The rectified, or d.c., voltage signal is used in conjunction with a desired or set speed signal to operate a hydraulic unit. This unit will then move the fuel controls in the appropriate direction to control the engine speed.

GOVERNORS AND OVER-SPEED TRIPS

These must be fully operational and regularly tested in accordance with manufacturers' instructions. Attention is drawn to the testing of over speed trip and protection devices. The condition of the linkage coupling the engine's fuel pump actuating levers and the governor is also to be regularly examined. The governor cannot compensate for either seized fulcrum pins or excessive clearances.

3

Boilers

MARINE BOILER

A boiler is defined as "a closed vessel in which water or other liquid is heated, steam or vapor is generated, steam is superheated, or any combination thereof, under pressure or vacuum, for use external to itself, by the direct application of energy from the combustion of fuels, from electricity or nuclear energy."

Also included are fired units for heating or vaporizing liquids other than water where these units are separate from processing systems and are complete within themselves. This definition includes water heaters that exceed 200,000 Btu/hr heat input, 200 degrees Fahrenheit at the outlet, or 120 gallons nominal water containing capacity.

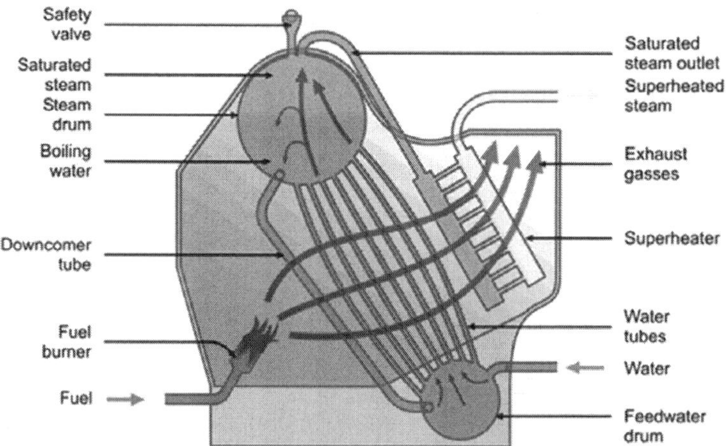

Boilers are one of the essensial equipments onboard ships. It's purpose is to provide heating to the the main deisel propulsion engine, to the bunker F.O. tanks(to make it less viscous for transfer purposes as well as easy ignition). The steam generated by the boiler can also be use for cleaning, and heating of the seawater in the freshwater generator thus evaporating it to make it potable water. There are two types of boiler mainly use in the Marine Industry, The water-tube boilers and the fire-tube boilers. The most common of which is the water-tube. In the water-tube type, water passes though lines of tube where it is heated by the used of a burner and a furnace.

BOILER

A portable boiler (preserved, Poland).

A stationary boiler (United States).

A boiler is a closed vessel in which water or other fluid is heated. The fluid does not necessarily boil. (In North America the term "furnace" is normally used if the purpose is not actually to boil the fluid.) The heated or vaporized fluid exits the boiler for use in various processes or heating applications, including water heating, central heating, boiler-based power generation, cooking, and sanitation.

MATERIALS

The pressure vessel of a boiler is usually made of steel (or alloy steel), or historically of wrought iron. Stainless steel, especially of the austenitic types, is not used in wetted parts of boilers due to corrosion and stress corrosion cracking. However, ferritic stainless steel is often used in superheater sections that will not be exposed to boiling water, and electrically-heated stainless steel shell boilers are allowed under the European "Pressure Equipment Directive" for production of steam for sterilizers and disinfectors.

In live steam models, copper or brass is often used because it is more easily fabricated in smaller size boilers. Historically, copper was often used for fireboxes (particularly forsteam locomotives), because of its better formability and higher thermal conductivity; however, in more recent times, the high price of copper often makes this an uneconomic choice and cheaper substitutes (such as steel) are used instead.

For much of the Victorian "age of steam", the only material used for boilermaking was the highest grade of wrought iron, with assembly by rivetting.

This iron was often obtained from specialist ironworks, such as at Cleator Moor (UK), noted for the high quality of their rolled plate and its suitability for high-reliability use in critical applications, such as high-pressure boilers.

In the 20th century, design practice instead moved towards the use of steel, which is stronger and cheaper, with welded construction, which is quicker and requires less labour. It should be noted, however, that wrought iron boilers corrode far slower than their modern-day steel counterparts, and are less susceptible to localized pitting and stress-corrosion. This makes the longevity of older wrought-iron boilers far superior to those of welded steel boilers.

Cast iron may be used for the heating vessel of domestic water heaters. Although such heaters are usually termed "boilers" in some countries, their purpose is usually to produce hot water, not steam, and so they run at low pressure and try to avoid actual boiling. The brittleness of cast iron makes it impractical for high-pressure steam boilers.

ENERGY

The source of heat for a boiler is combustion of any of several fuels, such as wood, coal, oil, or natural gas. Electric steam boilers use resistance- or immersion-type heating elements.

Nuclear fission is also used as a heat source for generating steam, either directly (BWR) or, in most cases, in specialised heat exchangers called "steam generators" (PWR). Heat recovery steam generators (HRSGs) use the heat rejected from other processes such as gas turbine.

CONFIGURATIONS

Boilers can be classified into the following configurations:

- *Pot boiler* or *Haycock boiler*/Haystack boiler: a primitive "kettle" where a fire heats a partially filled water container from below. 18th century Haycock boilers generally produced and stored large volumes of very low-pressure steam, often hardly above that of the atmosphere. These could burn wood or most often, coal. Efficiency was very low.

- Flued boiler with one or two large flues—an early type or forerunner of fire-tube boiler.

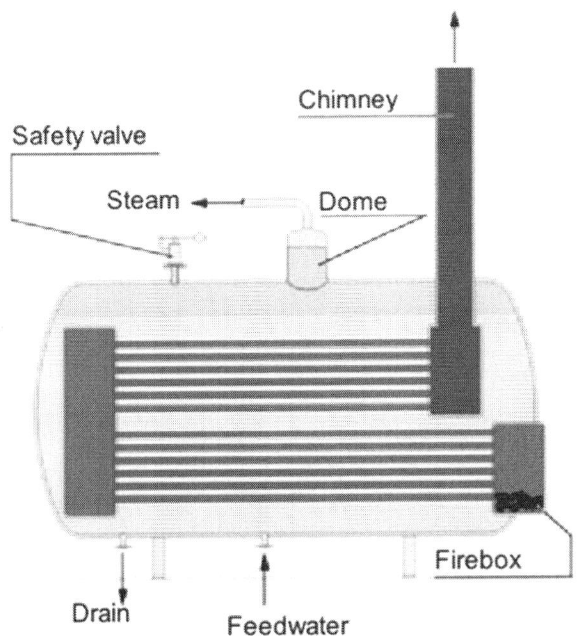

Diagram of a fire-tube boiler

Fire-tube boiler: Here, water partially fills a boiler barrel with a small volume left above to accommodate the steam (*steam space*). This is the type of boiler used in nearly all steam locomotives. The heat source is inside a furnace or *firebox* that has to be kept permanently surrounded by the water in order to maintain the temperature of the *heating surface* below the boiling point. The furnace can be situated at one end of a fire-tube which lengthens the path of the hot gases, thus augmenting the heating surface which can be further increased by making the gases reverse direction through a second parallel tube or a bundle of multiple tubes (two-pass or return flue boiler); alternatively the gases may be taken along the sides and then beneath the boiler through flues (3-pass boiler).

In case of a locomotive-type boiler, a boiler barrel extends from the firebox and the hot gases pass through a bundle of fire tubes inside the barrel which greatly increases the heating surface compared to a single tube and further improves heat transfer. Fire-tube boilers usually have a comparatively low rate of steam production, but high steam storage capacity. Fire-tube boilers mostly burn solid fuels, but are readily adaptable to those of the liquid or gas variety.

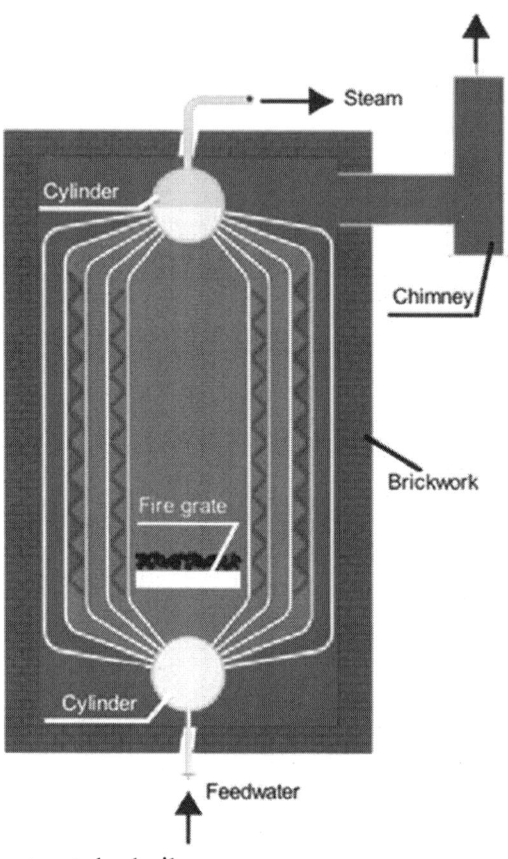

Diagram of a water-tube boiler.

Water-tube boiler: In this type, tubes filled with water are arranged inside a furnace in a number of possible configurations. Often the water tubes connect large drums, the lower ones containing water and the upper ones steam and water; in other cases, such as a mono-tube boiler, water is circulated by a pump through a succession of coils.

This type generally gives high steam production rates, but less storage capacity than the above. Water tube boilers can be designed to exploit any heat source and are generally preferred in high-pressure applications since the high-pressure water/steam is contained within small diameter pipes which can withstand the pressure with a thinner wall.

- Flash boiler: A flash boiler is a specialized type of water-tube boiler in which tubes are close together and water is pumped through them. A flash boiler differs from the type of mono-tube steam generator in which the tube is permanently filled with water. In a flash boiler, the tube is kept so hot that the water feed is quickly flashed into steam and superheated. Flash boilers had some use in automobiles in the 19th century and this use continued into the early 20th century..

1950s design steam locomotive boiler, from a Victorian Railways J class

- Fire-tube boiler with Water-tube firebox. Sometimes the two above types have been combined in the following manner: the firebox contains an assembly of water tubes, called thermic siphons. The gases then pass through a conventional firetube boiler. Water-tube fireboxes were installed in many Hungarian locomotives, but have met with little success in other countries.
- Sectional boiler. In a cast iron sectional boiler, sometimes called a "pork chop boiler" the water is contained inside cast iron sections. These sections are assembled on site to create the finished boiler.

SAFETY

To define and secure boilers safely, some professional specialized organizations such as the American Society of Mechanical Engineers(ASME) develop standards and regulation codes. For instance, the ASME Boiler and Pressure Vessel Code is a standard providing a wide range of rules and directives to ensure compliance of the boilers and other pressure vessels with safety, security and design standards. Historically, boilers were a source of many serious injuries and property destruction due to poorly understood engineering principles. Thin and brittle metal shells can rupture, while poorly welded or riveted seams could open up, leading to a violent eruption of the pressurized steam. When water is converted to steam it expands to over 1,000 times its original volume and travels down steam pipes at over 100 kilometres per hour. Because of this, steam is a great way of moving energy and heat around a site from a central boiler house to where it is needed, but without the right boiler feed water treatment, a steam-raising plant will suffer from scale formation and corrosion. At best, this increases energy costs and can lead to poor quality steam, reduced efficiency, shorter plant life and unreliable operation. At worst, it can lead to catastrophic failure and loss of life. Collapsed or dislodged boiler tubes can also spray scalding-hot steam and smoke out of the air intake and firing chute, injuring the firemen who load the coal into the fire chamber. Extremely large boilers providing hundreds of horsepower to operate factories can potentially demolish entire buildings.

A boiler that has a loss of feed water and is permitted to boil dry can be extremely dangerous. If feed water is then sent into the empty boiler, the small cascade of incoming water instantly boils on contact with the superheated metal shell and leads to a violent explosion that cannot be controlled even by safety steam valves. Draining of the boiler can also happen if a leak occurs in the steam supply lines that is larger than the make-up water supply could replace. The *Hartford Loop* was invented in 1919 by the Hartford Steam Boiler and Insurance Company as a method to help prevent this condition from occurring, and thereby reduce their insurance claims.

SUPERHEATED STEAM BOILERS

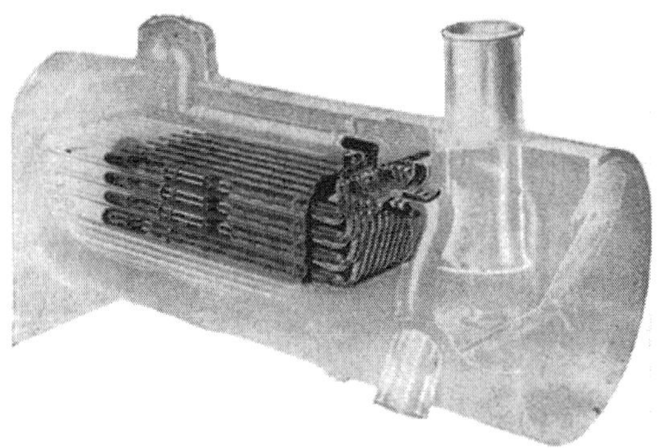

A superheated boiler on a steam locomotive.

Most boilers produce steam to be used at saturation temperature; that is, saturated steam. Superheated steam boilers vaporize the water and then further heat the steam in a *superheater*. This provides steam at much higher temperature, but can decrease the overall thermal efficiency of the steam generating plant because the higher steam temperature requires a higher flue gas exhaust temperature.

There are several ways to circumvent this problem, typically by providing an *economizer* that heats the feed water, a combustion air heater in the hot flue gas exhaust path, or both. There are advantages to superheated steam that may, and often will, increase overall efficiency of both steam generation and its utilization: gains in input temperature to a turbine should outweigh any cost in additional boiler complication and expense. There may also be practical limitations in using *wet* steam, as entrained condensation droplets will damage turbine blades.

Superheated steam presents unique safety concerns because, if any system component fails and allows steam to escape, the high pressure and temperature can cause serious, instantaneous harm to anyone in its path. Since the escaping steam will initially be completely superheated vapor, detection can be difficult, although the intense heat and sound from such a leak clearly indicates its presence.

Superheater operation is similar to that of the coils on an air conditioning unit, although for a different purpose. The steam piping is directed through the flue gas path in the boiler furnace. The temperature in this area is typically between 1,300 and 1,600 °C (2,372 and 2,912 °F). Some superheaters are radiant type; that is, they absorb heat by radiation. Others are convection type, absorbing heat from a fluid. Some are a combination of the two types. Through either method, the extreme heat in the flue gas path will also heat the superheater steam piping and the steam within. While the temperature of the steam in the superheater rises, the pressure of the steam does not and the pressure remains the same as that of the boiler. Almost all steam superheater system designs remove droplets entrained in the steam to prevent damage to the turbine blading and associated piping.

Supercritical steam generator

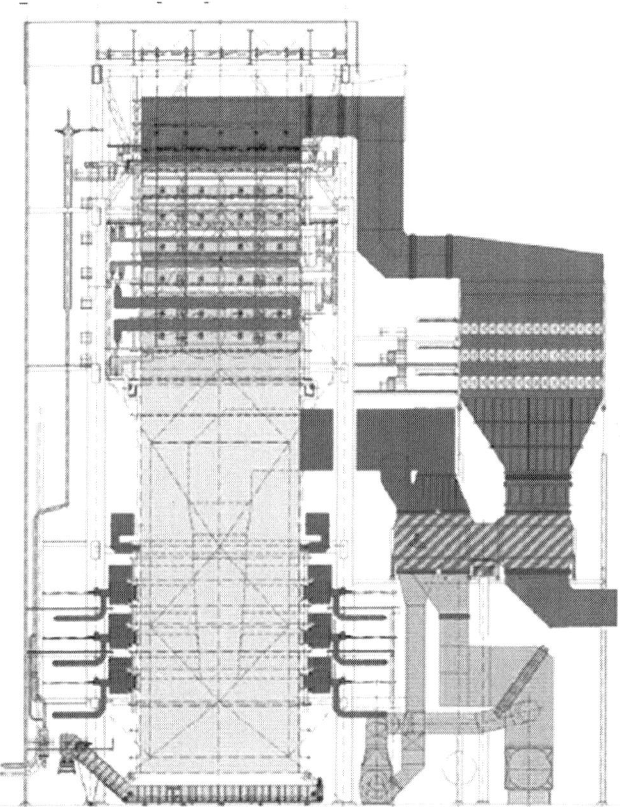

Boiler for a power plant.

Supercritical steam generators are frequently used for the production of electric power. They operate at supercritical pressure. In contrast to a "subcritical boiler", a supercritical steam generator operates at such a high pressure (over 3,200 psi or 22 MPa) that the physical turbulence that characterizes boiling ceases to occur; the fluid is neither liquid nor gas but a super-critical fluid. There is no generation of steam bubbles within the water, because the pressure is above the critical pressure point at which steam bubbles can form. As the fluid expands through the turbine stages, its thermodynamic state drops below the critical point as it does work turning the turbine which turns electrical generator from which power is ultimately extracted. The fluid at that point may be a mix of steam and liquid droplets as it passes into the condenser. This results in slightly less fuel use and therefore less greenhouse gasproduction. The term "boiler" should not be used for a supercritical pressure steam generator, as no "boiling" actually occurs in this device.

ACCESSORIES

Boiler fittings and accessories

- Pressuretrols to control the steam pressure in the boiler. Boilers generally have 2 or 3 pressuretrols: a manual-reset pressuretrol, which functions as a safety by setting the upper limit of steam pressure, the operating pressuretrol, which controls when the boiler fires to maintain pressure, and for boilers equipped with a modulating burner, a modulating pressuretrol which controls the amount of fire.
- Safety valve: It is used to relieve pressure and prevent possible explosion of a boiler.
- Water level indicators: They show the operator the level of fluid in the boiler, also known as a sight glass, water gauge or water column.
- Bottom blowdown valves: They provide a means for removing solid particulates that condense and lie on the bottom of a boiler. As the name implies, this valve is usually located directly on the bottom of the boiler, and is occasionally opened to use the pressure in the boiler to push these particulates out.
- Continuous blowdown valve: This allows a small quantity of water to escape continuously. Its purpose is to prevent the water in the boiler becoming saturated with dissolved salts. Saturation would lead to foaming and cause water droplets to be carried over with the steam – a condition known as priming. Blowdown is also often used to monitor the chemistry of the boiler water.
- Flash tank: High-pressure blowdown enters this vessel where the steam can 'flash' safely and be used in a low-pressure system or be vented to atmosphere while the ambient pressure blowdown flows to drain.
- Automatic blowdown/continuous heat recovery system: This system allows the boiler to blowdown only when makeup water is flowing to the boiler, thereby transferring the maximum amount of heat possible

from the blowdown to the makeup water. No flash tank is generally needed as the blowdown discharged is close to the temperature of the makeup water.

- Hand holes: They are steel plates installed in openings in "header" to allow for inspections & installation of tubes and inspection of internal surfaces.
- Steam drum internals, a series of screen, scrubber & cans (cyclone separators).
- Low-water cutoff: It is a mechanical means (usually a float switch) that is used to turn off the burner or shut off fuel to the boiler to prevent it from running once the water goes below a certain point. If a boiler is "dry-fired" (burned without water in it) it can cause rupture or catastrophic failure.
- Surface blowdown line: It provides a means for removing foam or other lightweight non-condensible substances that tend to float on top of the water inside the boiler.
- Circulating pump: It is designed to circulate water back to the boiler after it has expelled some of its heat.
- Feedwater check valve or clack valve: A non-return stop valve in the feedwater line. This may be fitted to the side of the boiler, just below the water level, or to the top of the boiler.
- Top feed: In this design for feedwater injection, the water is fed to the top of the boiler. This can reduce boiler fatigue caused by thermal stress. By spraying the feedwater over a series of trays the water is quickly heated and this can reduce limescale.
- Desuperheater tubes or bundles: A series of tubes or bundles of tubes in the water drum or the steam drum designed to cool superheated steam, in order to supply auxiliary equipment that does not need, or may be damaged by, dry steam.
- Chemical injection line: A connection to add chemicals for controlling feedwater pH.

Steam accessories

- Main steam stop valve:
- Steam traps:
- Main steam stop/check valve: It is used on multiple boiler installations.

Combustion accessories

- Fuel oil system:fuel oil heaters
- Gas system:
- Coal system:
- Soot blower

Other essential items

- Pressure gauges:
- Feed pumps:
- Fusible plug:
- Inspectors test pressure gauge attachment:
- Name plate:
- Registration plate:

Gas safe check

- It is essential to carry out gas safe check each year:

DRAUGHT

A fuel-heated boiler must provide air to oxidize its fuel. Early boilers provided this stream of air, or *draught*, through the natural action of convection in a chimney connected to the exhaust of the combustion chamber. Since the heated flue gas is less dense than the ambient air surrounding the boiler, the flue gas rises in the chimney, pulling denser, fresh air into the combustion chamber.

Most modern boilers depend on mechanical draught rather than natural draught. This is because natural draught is subject to outside air conditions and temperature of flue gases leaving the furnace, as well as the chimney height. All these factors make proper draught hard to attain and therefore make mechanical draught equipment much more reliable and economical.

Types of draught can also be divided into *induced draught*, where exhaust gases are pulled out of the boiler; *forced draught*, where fresh air is pushed into the boiler; and*balanced draught*, where both effects are employed. Natural draught through the use of a chimney is a type of induced draught; mechanical draught can be induced, forced or balanced.

There are two types of mechanical induced draught. The first is through use of a steam jet. The steam jet oriented in the direction of flue gas flow induces flue gasses into the stack and allows for a greater flue gas velocity increasing the overall draught in the furnace. This method was common on steam driven locomotives which could not have tall chimneys. The second method is by simply using an induced draught fan (ID fan) which removes flue gases from the furnace and forces the exhaust gas up the stack. Almost all induced draught furnaces operate with a slightly negative pressure.

Mechanical forced draught is provided by means of a fan forcing air into the combustion chamber. Air is often passed through an air heater; which, as the name suggests, heats the air going into the furnace in order to increase the overall efficiency of the boiler. Dampers are used to control the quantity of air admitted to the furnace. Forced draught furnaces usually have a positive pressure.

Balanced draught is obtained through use of both induced and forced draught. This is more common with larger boilers where the flue gases have to travel a long distance through many boiler passes. The induced draught fan works in conjunction with the forced draught fan allowing the furnace pressure to be maintained slightly below atmospheric.

BOILER (POWER GENERATION)

A boiler or steam generator is a device used to create steam by applying heat energy to water. Although the definitions are somewhat flexible, it can be said that older steam generators were commonly termed *boilers* and worked at low to medium pressure (1–300 psi or 6.895–2,068.427 kPa) but, at pressures above this, it is more usual to speak of a *steam generator*.

A boiler or steam generator is used wherever a source of steam is required. The form and size depends on the application: mobile steam engines such as steam locomotives, portable engines and steam-powered road vehicles typically use a smaller boiler that forms an integral part of the vehicle; stationary steam engines, industrial installations and power stations will usually have a larger separate steam generating facility connected to the point-of-use by piping. A notable exception is the steam-powered fireless locomotive, where separately-generated steam is transferred to a receiver (tank) on the locomotive.

STEAM GENERATOR (COMPONENT OF PRIME MOVER)

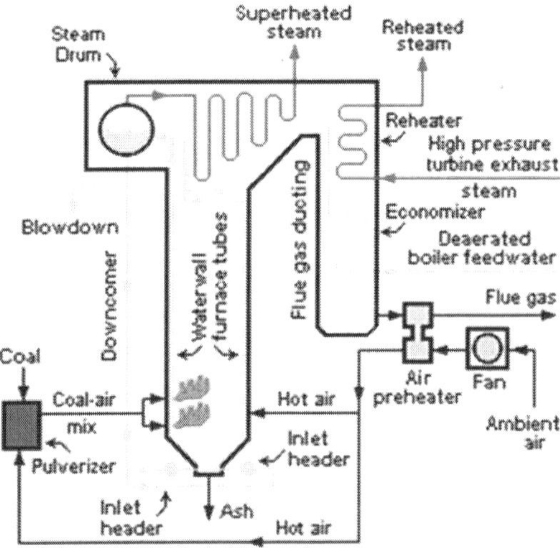

Fig. Type of Steam generator unit used in coal-fired power plants

The steam generator or boiler is an integral component of a steam engine when considered as a prime mover. However it needs be treated separately, as to some extent a variety of generator types can be combined with a variety of engine units. A boiler incorporates a firebox or furnace in order to burn the fuel and generate heat. The generated heat is transferred to water to make steam, the process of boiling. This produces saturated steam at a rate which can vary according to the pressure above the boiling water. The higher the furnace temperature, the faster the steam production. The saturated steam thus produced can then either be used immediately to produce power via a turbine and alternator, or else may be furthersuperheated to a higher temperature; this notably reduces suspended water content making a given volume of steam produce more work and creates a greater temperature gradient, which helps reduce the potential to form condensation. Any remaining heat in the combustion gases can then either be evacuated or made to pass through an economiser, the role of which is to warm the feed water before it reaches the boiler.

BOILER TYPES

Haycock and wagon top boilers

For the first Newcomen engine of 1712, the boiler was little more than large brewer's kettle installed beneath the power cylinder. Because the engine's power was derived from the vacuum produced by condensation of the steam, the requirement was for large volumes of steam at very low pressure hardly more than 1 psi (6.9 kPa) The whole boiler was set into brickworkwhich retained some heat. A voluminous coal fire was lit on a grate beneath the slightly dished pan which gave a very small heating surface; there was therefore a great deal of heat wasted up the chimney.

In later models, notably by John Smeaton, heating surface was considerably increased by making the gases heat the boiler sides, passing through a flue. Smeaton further lengthened the path of the gases by means of a spiral labyrinth flue beneath the boiler. These under-fired boilers were used in various forms throughout the 18th Century. Some were of round section (haycock). A longer version on a rectangular plan was developed around 1775 by Boulton and Watt (wagon top boiler). This is what is today known as a three-pass boiler, the fire heating the underside, the gases then passing through a central square-section tubular flue and finally around the boiler sides.

Cylindrical fire-tube boiler

An early proponent of the cylindrical form was the British engineer John Blakey, who proposed his design in 1774.Another early proponent was the American engineer, Oliver Evans, who rightly recognised that the cylindrical form was the best from the point of view of mechanical resistance and towards the end of the 18th Century began to incorporate it into his projects. Probably inspired by the writings on Leupold's "high-pressure" engine scheme that appeared in encyclopaedic works from 1725, Evans favoured "strong steam" i.e. non condensing engines in which the steam pressure alone drove the piston and was then exhausted to atmosphere. The advantage of strong steam as he saw it was that more work could be done by smaller volumes of steam; this enabled all the components to be reduced in size and engines could be adapted to transport and small installations. To this end he developed a long cylindrical wrought iron horizontal boiler into which was incorporated a single fire tube, at one end of which was placed the fire grate. The gas flow was then reversed into a passage or flue beneath the boiler barrel, then divided to return through side flues to join again at the chimney (Columbian engine boiler). Evans incorporated his cylindrical boiler into several engines, both stationary and mobile. Due to space and weight considerations the latter were one-pass exhausting directly from fire tube to chimney. Another proponent of "strong steam" at that time was the Cornishman,Richard Trevithick. His boilers worked at 40–50 psi (276–345 kPa) and were at first of hemispherical then cylindrical form. From 1804 onwards Trevithick produced a small two-pass or return flue boiler for semi-portable and locomotive engines. The Cornish boiler developed around 1812 by Richard Trevithick was both stronger and more efficient than the simple boilers which preceded it. It consisted of a cylindrical water tank around 27 feet (8.2 m) long and 7 feet (2.1 m) in diameter, and had a coal fire grate placed at one end of a single cylindrical tube about three feet wide which passed longitudinally inside the tank. The fire was tended from one end and the hot gases from it travelled along the tube and out of the other end, to be circulated back along flues running along the outside then a third time beneath the boiler barrel before being expelled into a chimney. This was later improved upon by another 3-pass boiler, the Lancashire boiler which had a pair of furnaces in separate tubes side-by-side. This was an important improvement since each furnace could be stoked at different times, allowing one to be cleaned while the other was operating.

Railway locomotive boilers were usually of the 1-pass type, although in early days, 2-pass "return flue" boilers were common, especially with locomotives built by Timothy Hackworth.

Multi-tube boilers

A significant step forward came in France in 1828 when Marc Seguin devised a two-pass boiler of which the second pass was formed by a bundle of multiple tubes. A similar design with natural induction used for marine purposes was the popular Scotch marine boiler.

Prior to the Rainhill trials of 1829 Henry Booth, treasurer of the Liverpool and Manchester Railway suggested to George Stephenson, a scheme for a multi-tube one-pass horizontal boiler made up of two units: a firebox surrounded by water spaces and a boiler barrel consisting of two telescopic rings inside which were mounted 25 copper tubes; the tube bundle occupied much of the water space in the barrel and vastly improved heat transfer. Old George immediately communicated the scheme to his son Robert and this was the boiler used on Stephenson's Rocket, outright winner of the trial. The design formed the basis for all subsequent Stephensonian-built locomotives, being immediately taken up by other constructors; this pattern of fire-tube boiler has been built ever since.

STRUCTURAL RESISTANCE

The 1712 boiler was assembled from riveted copper plates with a domed top made of lead in the first examples. Later boilers were made of small wrought iron plates riveted together. The problem was producing big enough plates, so that even pressures of around 50 psi (344.7 kPa) were not absolutely safe, nor was the cast iron hemispherical boiler initially used by Richard Trevithick. This construction with small plates persisted until the 1820s, when larger plates became feasible and could be rolled into a cylindrical form with just one butt-jointed seam reinforced by a gusset; Timothy Hackworth's *Sans Pareil 11* of 1849 had a longitudinal welded seam. Welded construction for locomotive boilers was extremely slow to take hold.

Once-through monotubular water tube boilers as used by Doble, Lamont and Pritchard are capable of withstanding considerable pressure and of releasing it without danger of explosion.

COMBUSTION

The source of heat for a boiler is combustion of any of several fuels, such as wood, coal, oil, or natural gas. Nuclear fission is also used as a heat source for generating steam.Heat recovery steam generators (HRSGs) use the heat rejected from other processes such as gas turbines.

Solid fuel firing

In order to create optimum burning characteristics of the fire, air needs to be supplied both through the grate, and above the fire. Most boilers now depend on mechanical draftequipment rather than natural draught. This is because natural draught is subject to outside air conditions and temperature of flue gases leaving the furnace, as well as chimney height. All these factors make effective draught hard to attain and therefore make mechanical draught equipment much more economical. There are three types of mechanical draught:

1. *Induced draught:* This is obtained one of three ways, the first being the "stack effect" of a heated chimney, in which the flue gas is less dense than the ambient air surrounding the boiler. The denser column of ambient air forces combustion air into and through the boiler. The second method is through use of a steam jet. The steam jet or ejector oriented in the direction of flue gas flow induces flue gases into the stack and allows for a greater flue gas velocity increasing the overall

draught in the furnace. This method was common on steam driven locomotives which could not have tall chimneys. The third method is by simply using an induced draught fan (ID fan) which sucks flue gases out of the furnace and up the stack. Almost all induced draught furnaces have a negative pressure.

2. *Forced draught:* Draught is obtained by forcing air into the furnace by means of a fan (FD fan) and duct-work. Air is often passed through an air heater; which, as the name suggests, heats the air going into the furnace in order to increase the overall efficiency of the boiler. Dampers are used to control the quantity of air admitted to the furnace. Forced draught furnaces usually have a positive pressure.
3. *Balanced draught:* Balanced draught is obtained through use of both induced and forced draft. This is more common with larger boilers where the flue gases have to travel a long distance through many boiler passes. The induced draft fan works in conjunction with the forced draft fan allowing the furnace pressure to be maintained slightly below atmospheric.

Firetube boiler

The next stage in the process is to boil water and make steam. The goal is to make the heat flow as completely as possible from the heat source to the water. The water is confined in a restricted space heated by the fire. The steam produced has lower density than the water and therefore will accumulate at the highest level in the vessel; its temperature will remain at boiling point and will only increase as pressure increases. Steam in this state (in equilibrium with the liquid water which is being evaporated within the boiler) is named "saturated steam". For example, saturated steam at atmospheric pressure boils at 100 °C (212 °F). Saturated steam taken from the boiler may contain entrained water droplets, however a well designed boiler will supply virtually "dry" saturated steam, with very little entrained water. Continued heating of the saturated steam will bring the steam to a "superheated" state, where the steam is heated to a temperature above the saturation temperature, and no liquid water can exist under this condition. Most reciprocating steam engines of the 19th century used saturated steam, however modern steam power plants universally use superheated steam which allows higher steam cycleefficiency.

Superheater

L.D. Porta gives the following equation determining the efficiency of a steam locomotive, applicable to steam engines of all kinds: power (kW) = steam Production (kg h^{-1})/Specific steam consumption (kg/kW h).

A greater quantity of steam can be generated from a given quantity of water by superheating it. As the fire is burning at a much higher temperature than the saturated steam it produces, far more heat can be transferred to the once-formed steam by superheating it and turning the water droplets suspended therein into more steam and greatly reducing water consumption.

The superheater works like coils on an air conditioning unit, however to a different end. The steam piping (with steam flowing through it) is directed through the flue gas path in the boiler furnace. This area typically is between 1,300–1,600 °C (2,372–2,912 °F). Some superheaters are radiant type (absorb heat by thermal radiation), others are convection type (absorb heat via a fluid i.e. gas) and some are a combination of the two. So whether by convection or radiation the extreme heat in the boiler furnace/flue gas path will also heat the superheater steam piping and the steam within as well. While the temperature of the steam in the superheater is raised, the pressure of the steam is not: the turbine or moving pistons offer a "continuously expanding space" and the pressure remains the same as that of the boiler. The process of superheating steam is most importantly designed to remove all droplets entrained in the steam to prevent damage to the turbine blading and/or associated piping. Superheating the steam expands the volume of steam, which allows a given quantity (by weight) of steam to generate more power.

When the totality of the droplets is eliminated, the steam is said to be in a superheated state.

In a Stephensonian firetube locomotive boiler, this entails routing the saturated steam through small diameter pipes suspended inside large diameter firetubes putting them in contact with the hot gases exiting the firebox; the saturated steam flows backwards from the wet header towards the firebox, then forwards again to the dry header. Superheating only began to be generally adopted for locomotives around the year 1900 due to problems of overheating of and lubrication of the moving parts in the cylinders and steam chests. Many firetube boilers heat water until it boils, and then the steam is used at saturation temperature in other words the temperature of the boiling point of water at a given pressure (saturated steam); this still contains a large proportion of water in suspension. Saturated steam can and has been directly used by an engine, but as the suspended water cannot expand and do work and work implies temperature drop, much of the working fluid is wasted along with the fuel expended to produce it.

Water tube boiler

Another way to rapidly produce steam is to feed the water under pressure into a tube or tubes surrounded by the combustion gases. The earliest example of this was developed by Goldsworthy Gurney in the late 1820s for use in steam road carriages. This boiler was ultra-compact and light in weight and this arrangement has since become the norm for marine and stationary applications. The tubes frequently have a large number of bends and sometimes fins to maximize the surface area. This type of boiler is generally preferred in high pressure applications since the high pressure water/steam is contained within narrow pipes which can contain the pressure with a thinner wall. It can however be susceptible to damage by vibration in surface transport appliances. In a cast iron sectional boiler, sometimes called a "pork chop boiler" the water is contained inside cast iron sections. These sections are mechanically assembled on site to create the finished boiler.

Supercritical steam generator

Supercritical steam generators are frequently used for the production of electric power. They operate at supercritical pressure. In contrast to a "subcritical boiler", a supercritical steam generator operates at such a high pressure (over 3,200 psi or 22.06 MPa) that actual boiling ceases to occur, the boiler has no liquid water - steam separation. There is no generation of steam bubbles within the water, because the pressure is above the critical pressure at which steam bubbles can form.

It passes below the critical point as it does work in a high pressure turbine and enters the generator's condenser. This results in slightly less fuel use and therefore less greenhouse gasproduction. The term "boiler" should not be used for a supercritical pressure steam generator, as no "boiling" actually occurs in this device.

WATER TREATMENT

Large cation/anion ion exchangersused in demineralization of boiler feedwater.

Feed water for boilers needs to be as pure as possible with a minimum of suspended solids and dissolved impurities which causecorrosion, foaming and water carryover. Various chemical treatments have been employed over the years, the most successful being Porta treatment. This contains a foam modifier that acts as a filtering blanket on the surface of the water that considerably purifies steam quality.

The most common options for demineralization of boiler feedwater are reverse osmosis (RO) and ion exchange (IX).

BOILER SAFETY

When water is converted to steam it expands in volume over 1,000 times and travels down steam pipes at over 100 kilometres/hr. Because of this, steam is a good way of moving energy and heat around a site from a central boiler house to where it is needed, but without the right boiler feed water treatment, a steam-raising plant will suffer from scale formation and corrosion. At best, this increases energy costs and can lead to poor quality steam, reduced efficiency, shorter plant life and an operation which is unreliable. At worst, it can lead to catastrophic failure and loss of life. While variations in standards may exist in different countries, stringent legal, testing, training and certification is applied to try to minimise or prevent such occurrences. Failure modes include:

- Overpressurisation of the boiler
- Insufficient water in the boiler causing overheating and vessel failure
- Pressure vessel failure of the boiler due to inadequate construction or maintenance.

Doble boiler

The Doble steam car uses a once-through type contra-flow generator, consisting of a continuous tube. The fire here is on top of the coil instead of underneath. Water is pumped into the tube at the bottom and the steam is drawn off at the top. This means that every particle of water and steam must necessarily pass through every part of the generator causing an intense circulation which prevents any sediment or scale from forming on the inside of the tube. Water enters the bottom of this tube at the flow rate of 600 feet (183 m) a second with less than two quarts of water in the tube at any one time.

As the hot gases pass down between the coils, they gradually cool, as the heat is being absorbed by the water. The last portion of the generator with which the gases come into contact remains the cold incoming water. The fire is positively cut off when the pressure reaches a pre-determined point, usually set at 750 psi (5.2 MPa), cold water pressure; asafety valve set at 1,200 lb (544 kg) provides added protection. The fire is automatically cut off by temperature as well as pressure, so in case the boiler were completely dry it would be impossible to damage the coil as the fire would be automatically cut off by the temperature.

Similar forced circulation generators, such as the Pritchard and Lamont and Velox boilers present the same advantages.

ESSENTIAL BOILER FITTINGS

- Safety valve
- Pressure measurement
- Blowdown Valves
- Main steam Stop Valve

- Feed check valves
- Fusible Plug
- Water gauge
- Low-Water Alarm
- Low Water Fuel Cut-out
- Inspector's Test Pressure Gauge Attachment
- Name Plate
- Registration Plate
- Boiler feedwater pump

Boiler fittings

- Safety valve: used to relieve pressure and prevent possible explosion of a boiler. As originally devised by Denis Papin it was a dead weight on the end of an arm that was lifted by excess steam pressure. This type of valve was used throughout the 19th century for stationary steam engines, however the vibrations of locomotive engines caused the valves to bounce and "fizzle" wasting steam. They were therefore replaced by various spring-loaded devices.
- Water column: to show the operator the level of fluid in the boiler, a water gauge or water column is provided
- Bottom blowdown valves
- Surface blowdown line
- Feed Pump(s)
- Circulating pump
- Check valve or clack valve: a non-return stop valve by which water enters the boiler.

STEAM ACCESSORIES

- Main steam stop valve
- Steam traps
- Main steam stop/Check valve used on multiple boiler installations

Combustion accessories

- Fuel oil system
- Gas system
- Coal system
- Automatic combustion systems

APPLICATION OF STEAM BOILERS

Steam boilers are used where steam and hot steam is needed. Hence, steam boilers are used as generators to produce electricity in the energy business. Besides many different application areas in the industry for example in heating systems or for cement production, steam boilers are used in agriculture as well for soil steaming.

TESTING STEAM GENERATORS

The preeminent code for testing fired steam generators is the American Society of Mechanical Engineers (ASME) performance test code, PTC 4. A related component is the regenerative air heater. A major revison to the performance test code for air heaters will be published in 2013. Copies of the draft are available for review. The European standards for acceptance test of steam boilers are EN 12952-15 and EN 12953-11. The British standards BS 845-1 and BS 845-2 remain also in use in the UK.

STEAM BOILER | WORKING PRINCIPLE AND TYPES OF BOILER

DEFINITION OF BOILER

Steam boiler or simply a boiler is basically a closed vessel into which water is heated until the water is converted into steam at required pressure. This is most basic definition of boiler.

WORKING PRINCIPLE OF BOILER

The basic working principle of boiler is very very simple and easy to understand. The boiler is essentially a closed vessel inside which water is stored. Fuel (generally coal) is bunt in a furnace and hot gasses are produced. These hot gasses come in contact with water vessel where the heat of these hot gases transfer to the water and consequently steam is produced in the boiler. Then this steam is piped to the turbine of thermal power plant. There are many different types of boiler utilized for different purposes like running a production unit, sanitizing some area, sterilizing equipment, to warm up the surroundings etc.

STEAM BOILER EFFICIENCY

The percentage of total heat exported by outlet steam in the total heat supplied by the fuel(coal) is called steam boiler efficiency.

$$Steam\ Boiler\ Efficiency(\%) = \frac{Heat\ exported\ by\ outlet\ steam}{Heat\ supplied\ by\ the\ fuel} \times 100$$

It includes with thermal efficiency, combustion efficiency & fuel to steam efficiency. Steam boiler efficiency depends upon the size of boiler used. A typical efficiency of steam boiler is 80% to 88%. Actually there are some losses occur like incomplete combustion, radiating loss occurs from steam boiler surrounding wall, defective combustion gas etc. Hence, efficiency of steam boiler gives this result.

TYPES OF BOILER

There are mainly two types of boiler – water tube boiler and fire tube boiler. In fire tube boiler, there are numbers of tubes through which hot gases are passed and water surrounds these tubes. Water tube boiler is reverse of the fire tube boiler. In water tube boiler the water is heated inside tubes and hot gasses surround these tubes. These are the main two types of boiler but each of the types can be sub divided into many which we will discuss later.

FIRE TUBE BOILER

As it indicated from the name, the fire tube boiler consists of numbers of tubes through which hot gasses are passed. These hot gas tubes are immersed into water, in a closed vessel. Actually in fire tube boiler one closed vessel or shell contains water, through which hot tubes are passed. These fire tubes or hot gas tubes heated up the water and convert the water into steam and the steam remains in same vessel. As the water and steam both are in same vessel a fire tube boiler cannot produce steam at very high pressure. Generally it can produce maximum 17.5 kg/cm^2 and with a capacity of 9 Metric Ton of steam per hour.

TYPES OF FIRE TUBE BOILER

There are different types of fire tube boiler likewise, external furnace and internal furnace fire tube boiler. External furnace boiler can be again categorized into three different types- 1) Horizontal Return Tubular Boiler. 2) Short Fire Box Boiler. 3) Compact Boiler. Again, internal furnace fire tube boiler has also two main categories such as horizontal tubular and vertical tubular fire tube boiler. Normally horizontal return fire tube boiler is used in thermal power plant of low capacity. It consists of a horizontal drum into which there are numbers of horizontal tubes. These tubes are submerged in water. The fuel (normally coal) burnt below these horizontal drum and the combustible gasses move to the rear from where they enter into fire tubes and travel towards the front into the smoke box. During this travel of gasses in tubes, they transfer their heat into the water and steam bubbles come up. As steam is produced, the pressure of the boiler developed, in that closed vessel.

ADVANTAGES OF FIRE TUBE BOILER

1. It is quite compact in construction.
2. Fluctuation of steam demand can be met easily.
3. It is also quite cheap.

DISADVANTAGES OF FIRE TUBE BOILER

1. As the water required for operation of the boiler is quite large, it requires long time for rising steam at desired pressure.

2. As the water and steam are in same vessel the very high pressure of steam is not possible.
3. The steam received from fire tube boiler is not very dry.

WATER TUBE BOILER

A water tube boiler is such kind of boiler where the water is heated inside tubes and the hot gasses surround them.

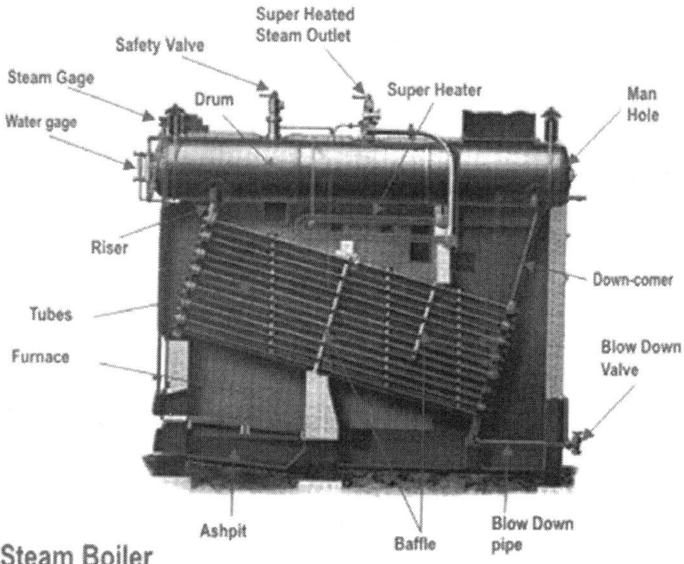

Steam Boiler

This is the basic definition of water tube boiler. Actually this boiler is just opposite of fire tube boiler where hot gasses are passed through tubes which are surrounded by water.

TYPES OF WATER TUBE BOILER

There are many types of water tube boilers, such as

1. Horizontal Straight Tube Boiler.
2. Bent Tube Boiler.
3. Cyclone Fired Boiler.

Horizontal Straight Tube Boiler again can be sub - divided into two different types,

- Longitudinal Drum Water Tube Boiler.
- Cross Drum Water Tube Boiler.

Bent Tube Boiler also can be sub divided into four different types,

- Two Drum Bent Tube Boiler.
- Three Drum Bent Tube Boiler.
- Low Head Three Drum Bent Tube Boiler.
- Four Drum Bent Tube Boiler.

ADVANTAGES OF WATER TUBE BOILER

There are many advantages of water tube boiler due to which these types of boiler are essentially used in large thermal power plant.

1. Larger heating surface can be achieved by using more numbers of water tubes.
2. Due to convectional flow, movement of water is much faster than that of fire tube boiler, hence rate of heat transfer is high which results into higher efficiency.
3. Very high pressure in order of 140 kg/cm^2 can be obtained smoothly.

DISADVANTAGES OF WATER TUBE BOILER

1. The main disadvantage of water tube boiler is that it is not compact in construction.
2. Its cost is not cheap.
3. Size is a difficulty for transportation and construction.

SCOTCH MARINE BOILER

German example. Note the steam dome, a typically German feature, and also the corrugated furnaces.

A "Scotch" marine boiler (or simply Scotch boiler) is a design of steam boiler best known for its use on ships.

The general layout is that of a squat horizontal cylinder. One or more large cylindrical furnaces are in the lower part of the boiler shell. Above this is a large number of small-diameter fire-tubes. Gases and smoke from the furnace pass to the back of the boiler, then return through the small tubes and up and out of the chimney. The ends of these multiple tubes are capped by a smokebox, outside the boiler shell.

The Scotch boiler is a fire-tube boiler, in that hot flue gases pass through tubes set within a tank of water. As such, it is a descendant of the earlier Lancashire boiler and like the Lancashire it uses multiple separate furnaces to give greater heating area for a given furnace capacity. It differs from the Lancashire in two aspects: a large number of small diameter tubes (typically 3 or 4 inches (76 or 102 mm) diameter each) are used to increase the ratio of heating area to cross-section. Secondly the overall length of the boiler is halved by folding the gas path back on itself.

COMBUSTION CHAMBER

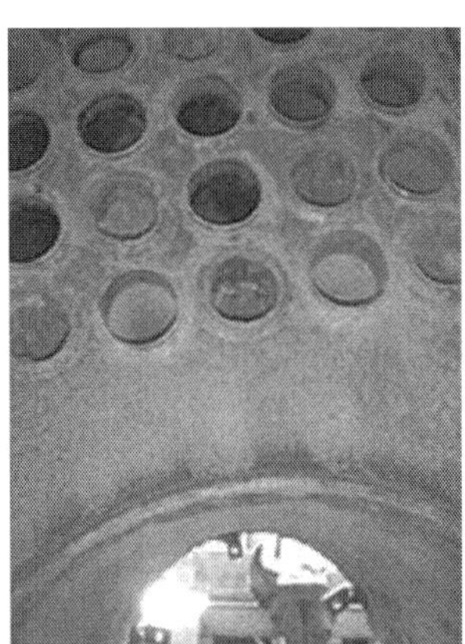

Inside the combustion chamber, looking up at the tubeplate

Rear face of the boiler of steam tug*Mayflower*, showing the stays supporting the combustion chambers

The far end of the furnace is an enclosed box called the *combustion chamber* which extends upwards to link up with the firetubes.

The front wall of the combustion chamber is supported against steam pressure by the tubes themselves. The rear face is stayed by rod stays through the rear shell of the boiler. Above the combustion chamber and tubes is an open steam collecting space. Larger long rod stays run the length of the boiler through this space, supporting the ends of the boiler shell.

With multiple furnaces, there is a separate combustion chamber for each furnace. A few small boilers did connect them into one chamber, but this design is weaker. A more serious problem is the risk of reversing the draught, where exhaust from one furnace could blow back and out of the adjacent one, injuring the stokers working in front of it.

Origins

The first recorded boiler of comparable form was used in a railway locomotive, Hackworth's *'Wilberforce'* class of 1830. This had a long cylindrical boiler shell similar to his earlier return-flued *'Royal George'*, but with the return flue replaced by a number of small firetubes, as had been demonstrated so effectively by Stephenson with his *'Rocket'* a year earlier. The novel feature of an entirely internal combustion chamber was used. Unlike the later Scotch boiler though, this was self-supported by its own stays, rather than using stays through the walls of the boiler shell. This allowed the entire assembly of outer tubeplate, furnace tube, combustion chamber and firetubes to all be removed from the boiler shell as one unit, simplifying manufacture and maintenance. Although a valuable feature, this became impractical for larger diameter chambers that would require the support of the shell.

VARIANTS

Number of furnaces

Typical practice for ships was to have two furnaces in each boiler. Smaller boilers might only have one, larger boilers commonly had three. The limitation in boiler size was the amount of work each stoker could do, firing one furnace per man. Larger ships (meaning anything above the smallest) would have many boilers. As with the Lancashire boiler, the furnace was often corrugated for strength. Various makers had their own particular ways of making these corrugations, leading to their classification for maintenance purposes under the broad titles of, Leeds, Morrison, Fox, Purves or Brown.

Wet back and dry back

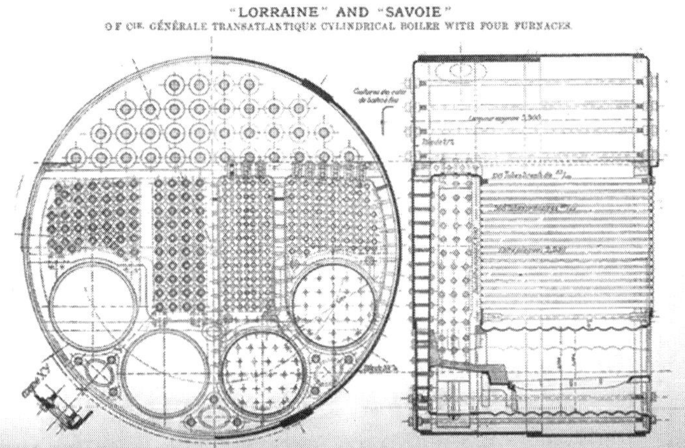

Cross and longitudinal section of a four-furnace boiler

The typical design is the "wet back", where the rear face of the combustion chamber is water-jacketed as a heating surface.

The "dry back" variation has the rear of the combustion chamber as an open box, backed or surrounded only by a sheetmetal jacket.This simplifies construction, but also loses much efficiency. It is only used for small boilers where capital cost outweighs fuel costs. Although the Scotch boiler is nowadays rarely the primary steam generator on a ship, small dry-back designs such as the Minipac are still encountered, for supporting secondary demands whilst alongside in port with the main boilers cold.

One interesting variant of the dry-back design has been a patent for burning ash-prone fuels. The rear of the combustion chamber is used as an access point for an ash separator, removing the ash before the small-diameter tubes.

Double-ended

The double-ended design places two boilers back-to-back, removing the rear wall of the boiler shell. The combustion chambers and firetubes remain separate. This design saves some structural weight, but it also makes the boiler longer and more difficult to install into a ship. For this reason they were not commonly used, although back-to-back arrangements of multiple single-ended boilers were common.

Inglis

The "Inglis" modification adds an extra combustion chamber where an additional single large flue returns from the rear to the front of the boiler. Flow through the multiple tubes is thus from front to back, and so the exhaust is at the rear. Multiple furnaces would share a single combustion chamber.

The major advantage of the Inglis is the extra heating area it adds, for a comparable shell volume, of perhaps 20%. Surprisingly this is not from the additional combustion chamber, but from lengthening the narrow firetubes. These can now run the full length of the boiler shell, rather than just the rather shorter distance from the inner combustion chamber to the front tubeplate. Despite this advantage, it is rarely used.

USE IN SHIPS

The Scotch marine boiler achieved near-universal use throughout the heyday of steam propulsion, particularly for the most highly developed piston engines such as the triple-expansion compounds. It lasted from the end of the low-pressure haystack boilers in the mid-19th century through to the early 20th century and the advent of steam turbines with high-pressure water-tube boilers such as theYarrow.

Large or fast ships could require a great many boilers. The *Titanic* had 29 boilers: 24 double-ended and 5 smaller single-ended. The larger boilers were

15 feet 9 inches (4.80 m) diameter and 20 feet (6.1 m) long, the smaller were 11 feet 9 inches (3.58 m) in length. All had three corrugated Morrison furnaces of 3 feet 9 inches (1.14 m) diameter, 159 furnaces in total, and a working pressure of 215 pounds per square inch (1,480 kPa).

Shipboard working examples

Numerous Scotch boilers are in use on ships as of 2011, and new boilers can be built to replace life-expired ones. Examples of preserved steam boats employing Scotch boilers include:

- Steam tug *Mayflower*, Bristol Industrial Museum

Mayflower's boiler was recently removed for restoration and could easily be seen in close-up.

- *Baltimore*, of 1906, at the Baltimore Museum of Industry, is the oldest working steam tug in the US.
- Steamship *Shieldhall* based in Southampton, UK, is fully operational and has two oil-fired Scotch boilers.
- Steam-powered icebreaker Stettin, operating on two coal-fired boilers
- Steam-powered drifter *Lydia Eva*, based in Lowestoft or Great Yarmouth, Fully operational coal-fired Scotch boiler.
- The steam tug *Kerne* built 1913. Operated out of Liverpool by the Steam Tug Kerne Preservation Society. One coal-fired wet back Scotch marine boiler with two furnaces. Max working pressure 180 pounds per square inch (1,200 kPa).
- The steam yacht *Louise* built 1902. Operating on Lake Geneva, Wisconsin by Gage Marine Corporation. One diesel-fired Scotch marine boiler powering a double-expansion steam engine, in excursion passenger service.
- The steam ship *Trafik* built 1892 in Stockholm Sweden by Bergsunds Mekaniska Verkstad as a passenger and freight ship on the Swedisk lake Vättern and with homeport Hjo. The machinery and its boiler are the original ones made for the ship in Stockholm.

PROCEDURE FOR STARTING AND STOPPING A BOILER

A boiler is one of those machineries that gets the ship going. A boiler is something, which though not required continuously in operating a ship, cannot be done away with. Moreover, it's a dangerous equipment which generates steam at extremely high pressure, and it is for this reason that proper care should be taken while operating it.

In this chapter we have brought to you a step-by-step procedure for starting and stopping a boiler on a ship. With this procedure you can never go wrong, as far as boilers are concerned. Starting and stopping a boiler was never so easy.

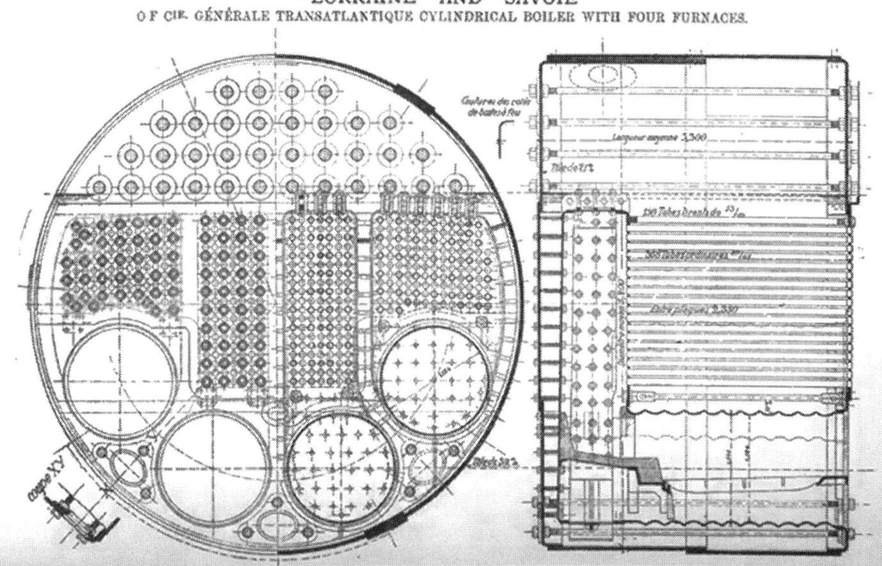

Starting a Boiler

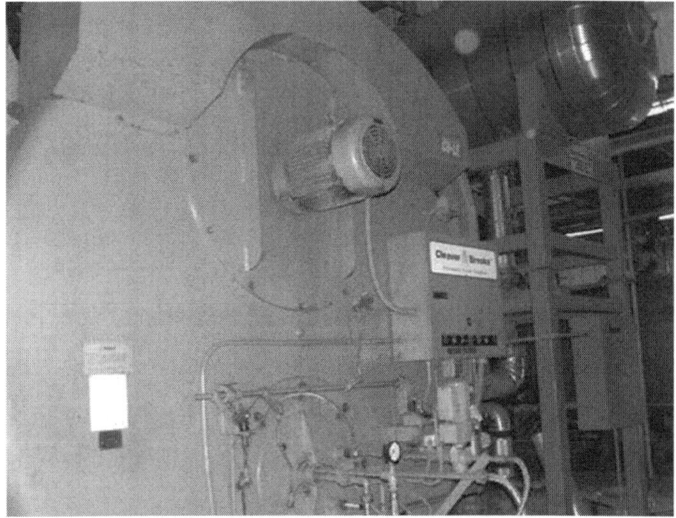

It is to note that the following steps may not apply to all types of boilers and each boiler requires some additional steps to be followed as per its system design. However, the basic steps remain the same:

1. Ensure that the vent valve on the boiler is open and check there is no pressure in the boiler.
2. Check that the steam stop valve is closed.
3. Check that all the valves for fuel are open, and let the fuel circulate through the system until it comes to the temperature required by the manufacturer recommendation.

4. Check and open the feed water valves to the boiler and fill the water inside the boiler drum to just above the low water level. This is done because it is not possible to start the boiler below the low water level due to safety feature which prevent boiler from starting. Also, the level is not filled much because if filled too much, the water inside the boiler might expand and over pressurize the boiler.
5. Start the boiler in automatic mode. The burner fan will start the purging cycle which will remove any gases present in the furnace by forcing it out through the funnel.
6. After the pre-set purge time the pilot burner will ignite. The pilot burner consists of two electrodes, through which a large current is passed, via the transformer, producing the spark between the electrodes. The pilot burner is supplied with diesel oil and when the oil passes over, the former ignites.
7. The main burner which is supplied by heavy oil catches fire with the help of pilot burner.
8. Check the combustion chamber from the sight glass to ensure the burner has lit and the flame is satisfactory.
9. Keep a close eye on the water level as the pressure increases and open the feed water when the level of water inside the gauge glass is stable.
10. Close the vent valve after the steam starts coming outside.
11. Open the steam stop valve.
12. Once the working steam pressure is reached, blow down the gauge glass and float chambers to check for the alarms.

Stopping a boiler

1. Ÿ If the boiler is needed to be stopped for longer duration for maintenance or opened up for survey, change the fuel to distillate fuel.
2. Ÿ If separate heating arrangement for heavy oil is present then there's is no need to change over to distillate fuel and the oil is kept on circulation mode.
3. Ÿ Stop the boiler automatic cycle.
4. Ÿ Close the steam stop valves.
5. Ÿ Close the boiler feed water valves.
6. Ÿ When the boiler pressure is just reduced to over atmospheric pressure the vent valve is kept open to prevent vacuum formation inside the boiler.

MARINE BOILERS STEAM AND WATER

In this marine steam boilers system, the circulation pump "J" runs continuously. The auxiliary steam boiler serves as steam separator for the entire system. The system is thus kept in hot condition and is ready to start instantly. If the steam from the exhaust boiler exceeds the steam demand onboard, a steam dump valve automatically dumps the surplus steam to a dump condenser.

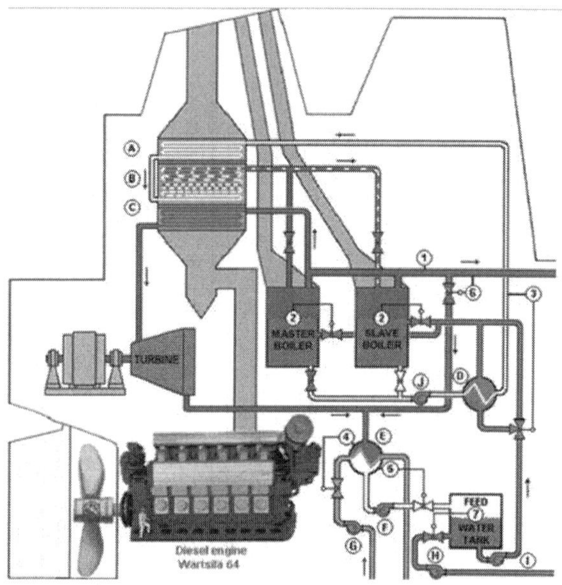

Fig. Steam flow diagram for an exhaust gas boiler and two oil-fired marine steam boilers

The energy content of the exhaust gases is utilized optimally in this marine boiler system. In addition to generating superheated steam for driving a turbo-alternator, the exhaust gas boiler also supplies steam for other requirements on board.

SOME THINGS TO THINK ABOUT

The outside temperature of the tubes of an Exhaust Gas Economizer of forced circulation type, the most common type today, is only a few °C higher than the water inside the tubes. The water temperature must therefore not be below 130°C if low temperature corrosionshall be avoided. Since ships mostly are equipped with open feed water systems where the feed water temperature is not higher than 50-80°C this feed water may be mixed with saturated water from the Oil Fired Boiler or from a steam separator drum, in proportions so that the water entering the EGE is minimum 130°C. As the feed water from the open feed water system contains oxygen, installation of a de-airator in this case is ecommended in order to avoid oxygen corrosion inside the tubes.

However, the most common way is to feed the EGE with saturated water from the OFB linked with EGE, since the temperature of the saturated water entering the EGE is reduced to 130°C in an externas heat changer by the feed water having a temperature of 50-80°C.

If the steam is used for production of electric energy by a turbo alternator, a superheater may be incorporated and located on the gas inlet side of the EGE.

Except for tankers that mostly are provided with large OFB's producing steam for the cargo pumps, the main part of the operating ships are furnished with an OFB and an EGE with capacities which are generally large enough to maintain only the bunker heating and the domestic heating.

However, where there are energy enough in the exhaust gases, it should be economically justified to installl a turbo-alternator set. When the EGB then is designed for a high rate of heat recovery from the gases, the EGB willl be larger in size and it willl thus have a larger water volume inside the tubes.

Due to increased steam production and water volume of the EGE, the OFB, steam receiver, has to be enlarged in order to obtain larger steam space and to withstand level variations which occur when the main engine starts and stops. When the main engine starts, and consequently the steam production takes place, the water in the EGB will expand and will be pressed from the EGE to the OFB where the water leivel increases. On the contrary, when the main engine stops, the steam production interrupts. The circulating water pump will then fill up the EGE with water that results in a decrease of the water level in the OFB. In case of a too small OFB compared with the steam production and the water content of the EGE, the water level variations in the OFB will be unacceptable and they will cause repeated, undesired high and low level alarms.

A. Economizer section of the Waste Heat Recovery Boiler
B. Preheats the circulating water before it enters the evaporator tubes. The counterflow principle is applied, i.e., the hottest exhaust gasses meet the heated feed water.
C. Evaporator section of the Waste Heat Recovery Boiler
D. Water evaporates and emulsion of steam and water flows back to the boiler. The counterflow principle is not applied since the evaporated steam rises in the tube bank and it would be disadvantageous to let the steam rise against the water flow.
E. Superheater section of the Waste Heat Recovery Boiler
F. Superheats the steam for the turbine. The counterflow principle is applied, i.e., the hottest exhaust gasses meet the superheated steam.
G. Heat exchanger
H. Preheats the boiler feedwater.
I. Condenser
J. The exhaust steam from the turbine and excess steam from the steam system condense and recycle.
K. Condensate pump
L. Cooling water pump
M. Seawater.
N. Make-up water pump

O. From softener unit.
P. Boiler feedwater pump
Q. Boiler water circulation pump
R. Alternator
S. Deliver electric power to the ships different el-consumers.
   1. Steam pressure control
   2. The pressure control loop adjusts the burner load according the steam demand.
   3. Water level control
   4. A simple control loop will do for a boiler with large amount of water and relatively small steam output. To minimize shrink and swell at start and stop of the burner it would be wise to have two setpoints for the water level. A lower level (abt. 40%) when the burner is stop and a higher (abt. 50%) when it's firing.
   5. Economizer inlet temperature control
   6. The feedwater is pre-heated in order to increase the efficiency of the plant. The circulating water to the exhaust gas boiler heats the feed water and the three-way valve on the inlet to the heat-exchanger controls the temperature. The economizer inlet temperature must never fall below 135°C to avoid corrosion on the economizer tubes.
   7. Condenser pressure control
   8. An absolute pressure transmitter and a controller adjust the cooling-water to the condenser to protect the condensate from being cooled down more than necessary.
   9. Condenser level control
   10. The level controller actuates the condensate outlet control valve.
   11. Steam dump control valve
   12. Takes care of excess steam from the waste heat boiler when the steam production exceeds the steam demand.
   13. Feedwater tank level control
   14. The level controller actuates the make up water control valve.

KEEPING A SLAVE BOILER PRESSURIZED IN A TWO BOILER SYSTEM

In a two boiler system it's often a problem to keep the slave boiler at operation pressure when the steam demand is low. This problem does not appear when the vessel is loading or unloading in a harbor since those operations normally need both boilers. On the other hand, when the ship is at sea and only one oil-fired boiler is used then the slave boiler tends to cool down far below the required stand by conditions. Different methods have been used to solve this problem. Installing steam heating coils in the bottom of the boiler is one method and a sophisticated start-and-stop method for the slave boiler's burner to keep the pressure at desired level is an other.

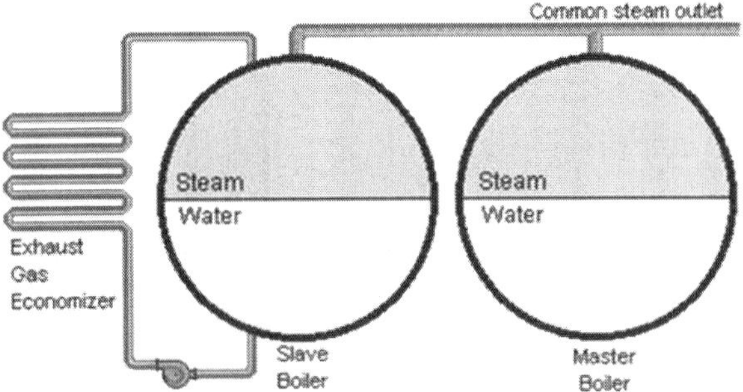

Those installations will be unnecessary if you happen to have an EGE, exhaust gas economizer.

Just connect the exhaust gas economizer to the slave boiler instead of the master boiler. This operation method will guarantee normal operation pressure on both boilers all the time at sea.

The method has been used in many ships and the chief engineers are satisfied with the result.

A two-boiler system has other advantages as well. It provides a fully automatic, very flexible and economic operation with maximum safety and availability ensured by the two separate units. At inert gas production, one boiler can be operated at a preset fixed output while the other will follow the load variations automatically.

When bringing a boiler on-line with another boiler, ensure that the working pressure is the same in both boilers and be sure to drain condensate from the steam line, before cracking the valve, to avoid water hammering. Let the steam line be heated up before the steam valve is fully opened.

OPERATION AND TYPES OF FIRE TUBE BOILER

Fire tube boiler is one of the most basic types of boiler and the design is also very old. It was popular in 18th century. It was mainly used for steam locomotive engines.

OPERATION OF FIRE TUBE BOILER

Operation of fire tube boiler is as simple as its construction. In fire tube boiler, the fuel is burnt inside a furnace. The hot gases produced in the furnace then passes through the fire tubes. The fire tubes are immersed in water inside the main vessel of the boiler. As the hot gases are passed through these tubes, the heat energy of the gasses is transferred to the water surrounds them. As a result steam is generated in the water and naturally comes up and is stored upon the water in the same vessel of fire tube boiler. This steam is then taken out from the steam outlet for utilizing for required purpose. The water is fed into the boiler through the feed water inlet. As the steam and water is stored is the same vessel, it is quite difficult to produce very high pressure steam from. General maximum capacity of this type of boiler is 17.5 kg/cm^2 and with a capacity of 9 Metric Ton of steam per hour. In a fire tube boiler, the main boiler vessel is under pressure, so if this vessel is burst there will be a possibility of major accident due to this explosion.

TYPES OF FIRE TUBE BOILER

According to the location of furnace there are two types of fire tube boiler and these are external furnace and internal furnace type. There are mainly three types of external furnace fire tube boiler. 1) Horizontal return tubular fire tube boiler. 2) Short fire box fire tube boiler. 3) Compact fire tube boiler. There are also two types of internal furnace fire tube boiler 1) Horizontal tubular. 2) Vertical tubular fire tube boiler.

WORKING PRINCIPLE OF HORIZONTAL RETURN FIRE TUBE BOILER

Horizontal return fire tube boiler is most suitable for low capacity thermal power plant. The main constructional features of this boiler are one big size steam drum which lies horizontally upon supporting structures. There are numbers of fire tubes come from furnace and also aligned horizontally inside the drum. When the drum is filled with water these tubes are submerged in water.

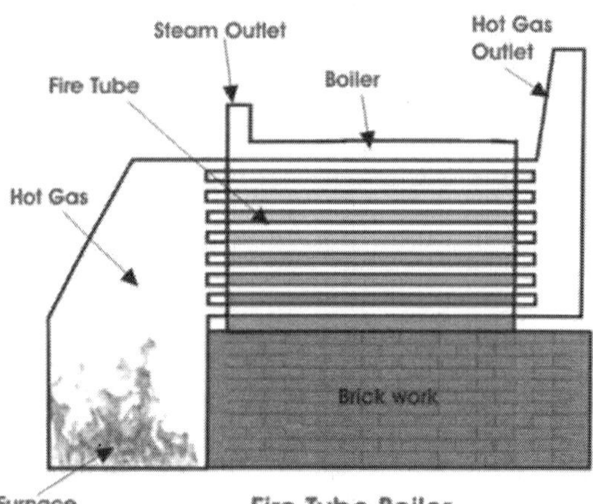

Fire Tube Boiler

The fuels (normally coal) burnt in the furnace and combustible gasses move into the fire tubes, travel through these tubes from rear to front of the boiler drum and finally the gases come into the smoke box. The hot gasses in the tubes under water transfer heat to the water via the tube walls. Due to this heat energy steam bubbles are created and come upon the water surface. As the amount of steam is increased in that closed drum, steam pressure inside the drum increases which increase significantly the boiling temperature of the water and hence rate of production of steam is reduced. In this way a fire tube boiler controls its own pressure. In other words this is a self pressure controlled boiler.

ADVANTAGES OF FIRE TUBE BOILER

1. Compact in construction.
2. Fluctuation of steam demand can be met easily.
3. Cheaper than water tube boiler.

DISADVANTAGES OF FIRE TUBE BOILER

1. Due to large water the required steam pressure rising time quite high.
2. Output steam pressure cannot be very high since the water and steam are kept in same vessel.
3. The steam received from fire tube boiler is not very dry.
4. In a fire tube boiler, the steam drum is always under pressure, so there may be a chance of huge explosion which resulting to severe accident.

STEAM PROPULSION

Conventional steam ship platforms are each configured in their own unique way to support the mission of the ship. Some ships have separate machinery spaces for steam generation and turbine operations, while others combine all main propulsion machinery in one machinery space. On ships using steam propulsion, boilers produce steam that is used by various components. The steam that goes to the main engines, ship's service turbine generators, boiler internal desuperheater and main feed pumps (on certain ships) is called main steam.

Steam generation in a conventional steam plant begins with the boiler. The D-type boiler has been installed in US Navy ships since 1950. Whether 600 psi or 1200 psi, D-type boiler construction is basically the same with a few exceptions, such as number of fuel oil burners and overall size and volume. Fuel oil burners are located on the boiler front and extend into the furnace to provide a means of firing the boiler. Depending on boiler design two to six burners are installed in the boiler.

The ship's main propulsion turbines are designed to efficiently convert the thermal energy of steam into useful mechanical energy to propel the ship through the water. As work is extracted from steam its pressure decreases. The high pressure turbinc is dcsigncd to efficiently extract work out of the high pressure steam as it initially enters the main propulsion turbines. The low pressure turbine is designed to efficiently extract work out of steam which is exhausting out of the high pressure turbine at a lower pressure. The propulsion turbines are designed to operate at high speeds. If the ship's propellers were to operate at such a high speed, cavitation around the propeller would occur and the ship would make little progress through the water. To allow both the turbine and the propeller to operate at their most efficient speeds, a main engine reduction gear is used to reduce the high rpms of the main propulsion turbines into a lower rpm for the main shaft.

Main propulsion boilers provide steam to the main propulsion turbines and auxiliary services in order to supply all shipboard steam systems in accordance with demand. It is designated as a D-type boiler because of the relative positions of the drums and side header which form the letter D. All D-type boilers are designated as uncontrolled superheat boilers because all the steam generated by the boiler must pass through the superheater. Superheater outlet temperature is a result of the combustion gas flow in proportion to the total amount of steam flow through all ranges (0 - 120%). The design characteristics ensure that the temperature will stabilize at set point.

The D-type boiler uses the principle of accelerated natural circulation to circulate water through the boiler. To enable this principle to work, relatively cool water will naturally circulate through large diameter pipes to distribution points low in the boiler. The downcomers are large diameter pipes connecting the steam drum with the water drum and lower headers to ensure proper circulation by delivering water from the steam drum to the water drum and lower headers. The downcomers are located between the inner and outer air casing to protect them from the direct radiant heat of the furnace.

The water drum is located at the bottom of the boiler below the main generating bank and acts as a lower reservoir of water for distribution to the main generating bank. Also, this large drum serves as a collection point for solids (sludge) that precipitate to the bottom that are removed by bottom blowdown.

The fuel oil service system provides the required amount of fuel oil to the boiler for operation. Its design allows operators to recirculate fuel for system testing and boiler light-off, redirect contaminated fuel, adjust for abnormalities in system pressure, and to quickly secure the system in case of a fuel leak or a casualty.

The combustion air system draws air from the outside atmosphere and directs it to the boiler to facilitate combustion. The proper amount of air to fuel is critical for the complete combustion of fuel. Too much or too little combustion air causes exhaust gases to become excessively white or black; situations that bring unique hazards to plant operation. The fireroom watch team must be able to monitor the exhaust gases to help maintain a clear smoke free stack. Smoke indicators and periscopes are installed to allow monitoring of the stack gases leaving the boiler. The smoke indicator is an electro-mechanical device and the periscope is an optical device. All ships have periscopes and many have electro-mechanical smoke indicators or stack gas analyzers. These devices are located above the economizer at the base of the stack so that combustion gases leaving the boiler must pass through its line of sight or the sensing element. From monitoring the stack gases, the combustion process can be adjusted for maximum efficiency or a casualty situation can be detected.

Automatic Boiler Controls (ABCs) are used extensively on modern surface ships to control the operation of the boiler and its auxiliaries under all load conditions from minimum to 120 percent. The ABCs include the various sub-systems necessary in maintaining both combustion rate and sufficient feed water supply to answer all steaming requirements within allowable tolerances.

The major piece of equipment which must be placed in operation in order to get the ship underway is the main engine. This requires placing the main condenser in operation and then testing and warming the main engine. If another boiler is needed to get underway, it should be lit-off early enough to ensure that it is on the line and the plant is stable prior to stationing the sea and anchor detail.

Before lighting fires in a boiler, there are several systems that must be checked and aligned so that electricity, atomizing air, combustion air, and fuel are available at the firing aisle. Power is needed for the electric forced draft blower, fuel oil service pump, and low pressure air compressor (LPAC). Low pressure air is required for automatic boiler control (ABC) systems and for fuel atomization. In addition, the boiler must have the proper level of water in the steam drum for light-off and the feed system must be tested for proper operation in preparation for supplying feedwater to the boiler after light-off. The following sections proceed step by step in the alignment and testing of these systems and ultimately for igniting the torch and establishing controlled combustion in the furnace.

BOILERS AND STEAM SYSTEMS

GENERAL DESCRIPTION

The steam generating plant consists of two auxiliary boilers and one exhaust gas economises Steam is required at sea for fuel, domestic water and cargo slop tank heating purposes. In port steam is used additionally for driving the power turbines of the cargo pumps and No. 1 water ballast pump. The steam demand of the plant, in port, is served by the boilers. At sea, steam demand is met by circulating boiler water from one of the auxiliary boilers through the exhaust gas economiscr, by one of the boiler water circulating pumps. The auxiliary boiler acts as a receiver for the steam generated by the economiscr.

The economiser is arranged in the main engine exhaust gas uptake to take waste heat from the main engine exhaust. An auxiliary boiler may be required at sea in low temperature areas, as well as reduced power operation of the main engine, such as during manoeuvring or slow steaming on passage when there will be insufficient waste heat to generate the required steam. Auxiliary Boiler

No. of sets:	2
Maker:	Hyundai Heavy Industries Ltd
Model:	HMT-50
Type:	Top fired rectangular water tube marine boiler
Evaporation:	50,000 kg/h
Steam Condition:	18 kg/cm2 saturated steam.
Fuel Oil:	HFO up to 700 cSt at 50°C
Safety Valve Setting:	20 kg/cm2
Fuel Oil Consumption:	3,850 kg/h at 100% evaporation

The furnace consists of gas-tight membrane walls, the downcomer pipes are located outside of the furnace. The fuel burner unit and associated combustion air inlet, is located in the roof of the furnace with the burner firing downwards using a steam assisted pressure jet burner.

At the furnace bottom, a refractory protects the furnace bottom from the combustion flame. Combustion gases flow downwards and through the lower part of the division tube wall and the lower section of the generating tube bank which connect the steam and water drums.

The gases then flow upwards on a return path through the upper part of the generating tube bank to the flue gas box at the top of the boiler. Radiant heat generates steam in the membrane furnace water wall tubes. The membrane wall has access doors to allow for furnace inspection and cleaning. The boiler structure is rigid enough to withstand rolling, pitching and shock loading of the ship operating in a seaway. The boiler is supported at the water drum and the water wall lower headers, and there are no rigid connections at any other points in order to allow for thermal expansion.

Furnace

Closely spaced water wall tubes of 76.2mm outside diameter, form the membrane walls at the side, roof, except for burner opening, rear, and front of the furnace. This construction is in order to increase the radiant heat absorption in the furnace and to make it strong enough to withstand vibration. The furnace is made completely gas-tight by the welded water wall construction. Situated at the top and bottom of the front and rear walls are water wall headers. Water enters the bottom headers and rises through the tubes to the top headers due to natural convection.

As the water rises, it is heated until its saturation temperature is reached and it then begins evaporating. This water- steam mixture is passed to the steam drum via the top headers.

Front and rear water wall tubes connect to steam and water headers at the top and bottom respectively; one end of each top header connects with the steam drum and one end of each bottom header connects with the water drum.

The roof, side and bottom water wall tubes are directly connected to the water and steam drums.

The steam generating bank of tubes, connecting steam and water drums, is located within the furnace.

Boiler Casing

As the furnace of the boiler is made completely gas-tight by the adoption of welded membrane water wall construction, no casing or refractory is required to contain the combustion gases. Mineral wool insulation is provided on the outer surface of the furnace water walls and this is covered by corrugated galvanised sheets to reduced heat transfer.

The maximum temperature on the casing surface will not exceed 60°C.

STEAM DRUM AND FITTINGS

The steam and water drums are fabricated using boiler steel plate of all welded construction. The steam drum has a horizontal perforated baffle plate covering the entire water surface in order to prevent droplets of water rising to the upper part of the steam drum.

A steam separator is provided to completely remove the moisture. The feedwater pipe enters the steam drum at the rear of the boiler and is attached to an internal perforated feed pipe which extends to the front of the steam drum.

This ensures that there is complete mixing of incoming feed with the existing boiler water and an equalising of temperatures. The chemical feedwater treatment pipe attaches to the internal feedwater pipe and this also ensures that there is complete mixing of the chemicals before the water reaches the downcomers.

The open ended surface blow off internal pipe extends to the surface of the steam drum to ensure that only floating solids on the water surface are discharged through this scum blowdown line. The boiler blowdown connection is fitted to the lower part of the water drum.

OPERATING PROCEDURES

Procedure for Preparing the Boiler for Service

The following steps should be taken before attempting to flash up the boiler.

a. All foreign materials must be removed from internal pressure parts.
b. All gas side-heating surfaces must be clean and all refractory be in good condition.
c. The furnace bottom and the burner wind box must been cleaned of oil and other debris.

d. All personnel not involved must remain clear of the boiler.
e. All manhole covers must be securely tightened.
f. Inspect safety valves and ensure that gags have been removed and easing levers are in good condition.
g. Open root valves for all instruments and controls connected to the boiler and check that they work as intended.
h. Open the vent valve of the steam drum.
i. Open all pressure gauge valves and check to ensure that all valves on the pressure gauge piping are open.
j. Check and close all blow-off valves and drain valves.
k. Fill the boiler until water level appears 25 to 50mm high in the gauge glasses.

RAISING PRESSURE WITH NO STEAM AVAILABLE FROM THE OTHER BOILER OR ECONOMISER

With the boiler water at the correct level and other checks made as above:

a. Set up the fuel system for diesel oil and circulate the fuel until all heavy fuel has been discharged from the fuel lines.
   Ideally the fuel system should have been flushed through with diesel oil prior to the previous shutdown.
b. Set the burner for air atomising, using an air pressure of 5 kg/cm2 and fuel pressure of 3 kg/cm2. Purge the furnace with the forced draught fan for one minute with vanes fully open.
c. Reduce the air pressure at the windbox to between 10 and 20mm WG and close recirculating valve.
d. Light the burner using the pilot burner and adjust air and fuel pressure to ensure stabilised combustion by using the furnace observation port and smoke indicator.
e. When raising the pressure, keep the burner firing for 5 minutes and out of service for 15 minutes repeatedly at the lowest fuel oil pressure (2.5kg/cm2) for one hour.
   Again, repeatedly light and shut down the burner to raise pressure as recommended on the pressure raising curve supplied by the manufacturer. A guideline would be to aim for lkg/cm2 after 2 hours firing, 5kg/cm2 after 2.75 hours firing and 12 kg/cm2 after 3.25 hours firing.
f. When the drum pressure has risen to about 2 kg/cm2, close the drum vent valve.
g. Drain and warm through all steam supply lines to ancillary equipment before putting the boiler on load.
h. Supply steam to one of the HFO service tanks.

When the tank is of sufficient temperature to be pumped by the HFO pump, supply steam to the HFO heater and prepare to change over from DO to HFO firing. The HFO must be thoroughly circulated through the system to ensure it is at the correct temperature for good combustion.

RAISING PRESSURE WITH STEAM AVAILABLE FROM THE OTHER BOILER OR ECONOMISER

a. Start the forced draught fan, open the inlet vanes and purge the furnace.
b. Ensure that the HFO system is correctly heated then start the HFO burning pump and circulate oil through the heater and burner manifold, open the recirculating valve and discharge the cold HFO in the line.
c. Reduce the air pressure at the windbox to between 10 and 20mmWG.
d. Close the recirculating valve.
e. Light the burner and adjust the air and fuel pressure to ensure stabilised combustion, using the furnace observation port and smoke indicator. Boiler pressure must be raised gradually over a period of hours in accordance with the manufacturer's instructions. The recommendations are the same as in item e) in the section; Raising Pressure With No Steam Available
f. When the drum pressure has risen to about 2 kg/cm2, close the drum vent valve.
g. Drain and warm through all steam supply lines to ancillary equipment before putting the boiler on load.

Shutting Down

a. Operate sootblowers before shutting down the boiler whenever possible.
b. Shut down the burner.
c. Continue operation of the forced draught fan for a short while after shutting down, keeping an air pressure of 150mm WG at burner inlet and purge the furnace of combustible gases.
d. Maintain the water level visible at about 50mm in the gauge glass and when the boiler is closed raise the water level 70mm to 120mm above the normal water level.
e. Open the drum vent valve when the boiler pressure reaches about 2 kg/cm2.
f. Change the fuel system to diesel oil and circulate back to the tank. (Note ! If steam is to remain available from the other boiler or economiser, the boiler HFO system should remain in use and there is no need to change to diesel oil.)
g. When fuel oil has been purged, shut down the fuel system.

After the boiler has been shut down for 4 hours the forced draught fan may be used to assist cooling down should immediate access be required.

However, to avoid the risk of damage to refractory, allow the boiler to cool down under natural means if possible.

INSPECTION CARRIED OUT IN BOILER ECONOMIZER

To obtain an acceptable degree of efficiency and reduce fuel consumption as much as possible by introducing further heat recovery surface so that the gas temperature at the funnel may be as low as practicable, the gas temperature leaving a boiler cannot be reduced much below 30°C above the saturation temperature. In radiant types a much higher exit gas temperature is usually found. To carry out this further heat exchange, surfaces such as economizers and air heaters are commonly used.

In many radiant boiler types, economizers are also found arranged integrally within the boiler unit. In this location they consist of a number of multi-loop elements of plain tubes connected at their ends to inlet and outlet headers.

Since are situated in a hot gas temperature zone and are required to perform a considerable heat exchange duty, a portion of the water pumped through them may be converted into steam. These steaming economizers are arranged so that water enters the lower header and the steam and water mixture leaves from the top header to the steam drum where the steam and water separate.

Economizers are used externally to boilers for further heat recovery. Economizers are found in the cooler gas zone and are fed with water temperatures around 116°C or 185°C depending upon whether the feed cycle includes high pressure feed heaters after the de-aerator.

BOILER WITH THE ECONOMIZER

INSPECTION ON GAS SIDE:

Before going into economizer inspection, first inspect the gas side of the boiler. It gives you a clear picture of boiler working condition and the efficiency of heat transfer surfaces.

1. Check exterior of drums for sign of tube roll, leakage, corrosion, soot erosion and overheating.
2. Condition of outside drum insulation.
3. Drum seals for signs of air leakage.
4. Inspect drum support for cracks and expansion clearance.
5. Check all the blow-down connection for expansion and flexibility of support.
6. Inspect all piping and valves for leaks.
7. Visually check water wall tubes and fins for cracks.
8. Check exterior of all tubes for corrosion, carbon-build up, erosion, blisters and sagging.
9. Inspect tubes at soot blower for sign of steam impingement.

10. Check header seals for signs of air leakage.
11. Examine exterior of headers for corrosion, erosion, thermal cracking and condition of insulation.
12. Condition of refractory.
13. Around the burner assembly check refractory, tube condition and accumulation of soot or carbon.
14. Check soot blowers for distortion, worn bearings, rubbing of tubes, condition of nozzle cracks, freedom of movement and effective lubrication.

INSPECTION ON ECONOMIZER:

1. The major problem at the economizer section is low temperature corrosion and problems from gas side deposits.
2. Sliding and leaky expansion joints at the casing may allow accumulation of soot with severe acid attack.
3. Inspection of tubes bends by opening the inspection covers needs to be carried out to check these.
4. Uptake area may show cracked expansion bellows sign of acid corrosion.

General cleanliness of these areas indicates the combustion performance in boiler. Safety valves are fitted to protect the boiler from the effect of over pressure. At least two safety valves are fitted to each boiler steam drum, but if there is a super heater, another safety valve should be fitted on it.

INTRODUCTION: MARINE BOILER SAFETY VALVES

The pressure setting of the superheater safety valve should be less that the designed pressure of the boiler, i.e. less than that of the steam drum safety valve, to ensure flow of steam through the superheater under blow off conditions. The pressure setting of one steam drum safety valve should be same as the design pressure of the boiler. The pressure setting of another safety valve should be 2-3 % more than the designed pressure of the boiler.

Classification of Boiler Safety Valves

There are three types of safety valves used in marine boilers:

1. Improved high lift safety valve
2. Full lift safety valve
3. Full bore safety valve

BOILER SAFETY VALVE

Improved High Lift Safety Valve:

1. Wingless valve improves steam flow and reduces risk of seizure.

2. Waste steam pressure acting on the piston gives increasing valve lift.
3. Special shaped seat deflects steam towards lip on valve and increases valve lift.
4. The valve lifts, the force to compress the spring increases, so the higher valve lifts the greater the increasing in boiler pressure.
5. Waste steam pressure keeps cylinder in place while piston moves, also by having a floating cylinder, seizure risk is reduced.
6. A lip is placed around the valve seat so that when the valve lid lifts, escaping steam is trapped in the annular space around the valve face, the resultant build–up of pressure acting upon the greater valve lid area causes the valve to lift sharply. This arrangement gives another advantage to close the valve cleanly and sharply with very little blow down effect.
7. The improved high lift safety valve makes use of waste steam pressure to increase the valve lift; this is done by allowing the pressure to act upon the lower spring carrier which fits within a floating ring so forming in effect a piston. The pressure acts upon this piston causing it to move up, helping to compress the spring and so increasing the valve lift.
8. Loose fitting key or pad lock is provided to ensure proper closing of valve.
9. Loose pin is provided to secure valve lid and allow thermal expansion.
10. Adjustment of the valve is carried out by means of a compression nut screwing down on to the top spring plate.
11. A compression ring is fitted after the final adjustment to ensure no further movement takes place.
12. A cap is then fitted over the compression nut and the top of the valve spindle, a cotter is passed through and padlocked to prevent tampering by unauthorized person.
13. Clearance between this cap, the valve spindle and cotter are such as to prevent the valve being held down externally.
14. Easing gear is fitted so that in the event of an emergency the valve can be opened by hand to a full lift ¼ D to release the boiler pressure.
    Valve Area: $As = A \times (1 + Ts/ 555)$
    - As- Aggregate area through the seating of valve (mm2) for superheated steam.
    - A-Aggregate area through the seating of valve (mm2) for saturated steam.
    - Ts- Degree of superheated steam in ^{o}C.
15. Valve Area (As) greater than (A) due to specific volume of steam increases with increases of temperature at constant pressure and more escape area is required to avoid accumulation of pressure.

16. The area of valve chest must be at least (1/2) A.
17. The waste steam pipe and steam passage must be at least 1.1× A.

Manual Hand Trying of Boiler Safety Relief Valve:

To check the proper working condition of the boiler safety valve we carry out the "Hand trying out the Boiler Safety valve" at regular intervals. The safety valve is provided with the easing gear which manually lifts the safety valve and releases the excess pressure in the boiler. When the easing gear is pulled, the valve will be opened by hand to a full lift of ¼ D to release the boiler pressure. Before carrying out the process the boiler safety valve has to be drained.

BOILER SAFETY VALVE DRAIN:

Draining of the boiler safety valve is necessary as to prevent any build-up of water in the pipe line causing head of water to form over the valve lid so increasing the blow off pressure. So at regular intervals the boiler safety valve should be drained.

1. Drain pipe must be fitted to the lowest part of the valve chest on the discharge side of the valve.
2. The pipe should be led clear of the boiler.
3. The pipe must have no valve or cock fitted through its length.
4. The open drain of the pipe should be regularly checked.
5. If the pipe becomes chocked, there is possibility of overloading the valve due to hydraulic head, or damage due to water hammer.
6. The waste steam pipe of the boiler safety valve should be well secured so that no load of the pipe is on the safety valve, which can be the cause of additional stress on the valve.

Pressure Setting of the Boiler Safety Valve:

If it is found that the boiler safety relief valve is not lifting at the designed lifting pressure, manual pressure setting of the boiler safety valve has to be done for the proper and safe operation of the boiler. The adjustment can be carried out on this type of valve to give the desired discharge and blow down characteristic.

1. Safety valve pressure setting can be done from high to low pressure or vice versa.
2. Take necessary personal safety precaution and arrange tools i.e. gagging tool and master gauges.
3. Slowly raise the boiler pressure and blow off the safety valves manually few times for thermal expansion and to reduce the thermal stress on the valves.
4. Then screw down all the safety valves higher than the setting pressure at which you are going to set.

5. Raise the boiler steam pressure 2-3 % more than the designed pressure of the boiler, then stop firing and unscrew the first valve slowly, when it blows off at 2-3 % more than the designed pressure then note this opening and closing pressure of the valve and finally gag it.
6. Raise the boiler pressure at the designed pressure of the boiler and unscrew the 2nd valve, when it blows off at designed pressure then note this opening pressure and check the closing pressure also. Recheck the setting pressure and gag the valve.
7. Then set the superheater safety valve lower than the designed pressure of the boiler in same procedure.
8. Finally take out the gagging tools. Pressure setting should be done in presence of surveyor

The boiler is vital equipment on ships. It is used as main propulsion (in steam ships) and for auxiliary heating in other ships. It is very sensitive and dangerous equipment, where there should be regular inspections and surveys carried out to avoid accidents and outages.

BOILER INSPECTION

Normally boiler inspection will be carried out onboard the ship by a port state control and during the dry dock. They are used to carry out the inspection and see the working condition of the boiler. During the inspection they will conduct an in-depth analysis of the boiler condition considering various factors to find the working condition of the boiler. If necessary they will replace damaged parts of the boiler needed for continued safe operation.

NEED FOR BOILER SURVEY OR INSPECTION

1. Boilers are inspected to maintain the Class requirement.
2. Regular internal inspection and external examination during such survey constitute the preventive maintenance schedule the boiler goes through to have a safe working condition.

FREQUENCY OF BOILER SURVEY

1. Water tube high pressure boilers are surveyed at two year intervals.
2. All other boilers, including exhaust gas boilers, are surveyed at two yearly intervals until they are eight years old and then surveyed annually.

PLANNING FOR BOILER SURVEY

1. Confirm time available, manpower, and time required.
2. Check before shutting down boiler.
3. Check for spares e.g. manhole door joints, gauge glass, packing and steam joints.

4. Check the tools required e.g. gagging tool, torque spanner, rope, chain block etc.
5. Check manual for special instruction and past records.
6. Steam requirement for the next port should be considered e.g. Tankers require steam in discharged Port.
7. Briefing to other engineers of work involved.

SHUTTING DOWN THE BOILER FOR INSPECTION

Before inspection is to be carried out, the boiler which is firing should be shut down. These are the steps to be followed before shutting down the boiler for inspection.

1. Inform the chief engineer and inform the duty officer in the bridge.
2. Change over M/E, A/E, and Boiler to diesel oil.
3. Top up diesel oil service tank, stop heavy oil and lube oil purifiers.
4. Stop all tank and tracing steam heating and carry out soot blowing.
5. Change over from automation to manual firing of boiler.
6. Stop the firing of the boiler and purge boiler for three to five minutes.
7. Switch off power and off the circuit breaker for forced draught fan, FO pump, feed pump, and combustion control panel. Hang necessary notices.
8. Shut main steam-stop valve and shut all fuel valves to boiler.
9. Let the boiler cool down, do not blow down now.
10. When the boiler pressure is about 4 bars, carry out blow down.
11. When boiler pressure is slightly higher than atmospheric pressure, open the vent cock to prevent formation of vacuum.
12. Let the boiler cool down.
13. Once sufficient cooled, open top manhole door first with all safety precaution.
14. Mark the nut on the top manhole, slacken the dog-nut, and secure it with a rope.
15. Knock the manhole door gently, but do not open it as it may contain steam or hot water.
16. Conform nothing coming out; open the door fully with the help of securing rope.
17. Do not open immediately open the bottom door, since the boiler is still hot and if opened relatively cool current of air will pass through the boiler causing a thermal shock.
18. Allow further cool down before opening bottom manhole door.
19. Open the bottom manhole door with the same precautions and open the furnace side door also.
20. Ventilate foe period of 12 to 24 hours.
21. Then check for oxygen, flammable vapour, and toxic gasses.
22. If it is safe, prepare for entry.

PREPARATION FOR ENTRY

These are the steps to be carried out before entering the boiler for inspection.

1. Prepare a long rope, wooden plank oxygen analyzer, safety hand lamp, and safety torch attached with rope.
2. Get a pouch to carry tools and keep track of the number of tools to be brought into boiler.
3. Personnel safety protection wear, e.g. helmet, safety shoes, hand gloves, etc.
4. No extra instruments to be brought in and clear pocket contents as it may fall into boiler.
5. Keep an emergency breathing apparatus ready.
6. Remain in communication and ensure proper lighting.
7. Check boiler internals before making an entry, e.g. foothold and handhold.

SUPER HEATERS

The superheater is a device which converts saturated steam or wet steam to dry steam, and it is used in driving the lager turbines in the marine propulsion system. In the superheating process the temperature of the steam is only raised, keeping the pressure at a constant level.

Superheating process can be done by three methods:

1. Radiant superheating: In this type, the superheating tubes are placed directly in the combustion chamber.
2. Convention superheating: In this type of super heaters the superheating tubes are placed outside the combustion chamber on the path of the hot gases.
3. Separately fired: In this type the superheater tubes are placed in the separate combustion chamber outside the boiler. This is separately fired to maintain the required temperature of the superheated steam outlet.

In the superheater zone the products of combustion were still at a high temperature and deposits from impurities in the fuel condensed out on the tubes, reducing heat transfer and steam temperature. Eventually gas passages between the tubes would become so badly blocked that the forced draught fans would be unable to supply sufficient air to the burners, combustion become impaired and the fouling condition accelerated.

Sodium and vanadium compounds present in the deposits proved very corrosive to superheater tube causing frequent repeated failure. Due to the fouled conditions there was a loss of efficiency and expensive time consuming cleaning routines were required.

INSPECTION ON SUPERHEATER

1. Internal and external examination of heaters.
2. Thermal crack at the headers due to high stresses set up across the thick welded section is possible.
3. Super heater safety valve and stop valve.
4. Super heater drains and vents valves and manhole openings to check.
5. Efficiency of the "screen" plates to ascertain –these protect headers from direct heat of furnace.
6. Superheater tubes are also prone to high temperature creep failures and thermal fatigue cracking sudden quenching can cause fatigue failure.
7. Check for deposit accumulation in header.
8. Drain valve from headers to examine.

SUPER HEATER WALK-IN SPACES:

1. Supports of horizontal super heater tubes to check for burning away and leave the unit unsupported and cause drainage problems.
2. Super heater support tubes may also crack due to effect of bending fatigue stresses due to misalignment of tubes in the tube holes.
3. Build-up of deposit is most troublesome defect in super heater. These may result in high furnace pressure, loss of super heater and poor combustion.
4. Special attention and suspicion to be reserved for tubes through which there still exist gas paths as they operate under excessive metal temperature.
5. Oxide scaling inside or outside may cause tube failure and worst case hydrogen fire when iron burns in steam at above 700*C in exothermic reaction, and destroys all boiler, economizer and air heater.

   Now you have a clear picture on the various inspections carried out on the marine boiler parts for the safe and efficient working of the boiler.

STEAM DRUM

HEADERS

Boiler headers are the water feeders to the generating tubes in boiler. The headers are connected in between the steam drum and the water drum. Normally the water from the water drum enters the main headers from there and many generating tubes are connected where the steam is generated.

Rear and Side Wall Headers

1. Sufficient doors or handhole plugs to remove for assessment of internal condition of headers and tubes.

2. Check for pitting and corrosion of headers, rear walls, floors, roofs, and side wall tubes.
3. Check for casing defects for possible gas or air leakage.

Bottom Header:

This contains the furnace tubes and the down comer tubes. A number of handhole doors is provided for internal inspection and repair to the tubes.

- Inspection for deposits of sludge must be carried out during the survey.
- Regular blowing down from this header will be necessary to keep it clear of sludge deposits.

REPAIRS IN SMOKE TUBE BOILERS

Procedure for Plugging of a Damaged/ Busted Smoke Tube:

1. Hydrostatic testing to mark the leaky tubes.
2. Cut the tubes on one end and clear of the tube plate. At the other end the tube is collapsed inside the tube plate.
3. Pull out the tube from the collapsed end.
4. Insert a short tube into the tube plate and weld it in place.
5. Lap the spare tapered plugs on both stud ends in the tube plates.
6. Insert the tube plugs and tack weld it.
7. Alternatively, the plugs can be held in place by a long steel bar threaded and bolted at both ends.
8. Hydrostatic pressure test to confirm no leaks.
9. Flush up the boiler and re-inspect the plugs for leaks under full steam pressure.

Temporary Repairing Procedure to Rectify the Leakage in Smoke Tube:

1. Stop the burner, allow the boiler to cool and remove the soot.
2. Allow boiler to depressurize, and open the blow down valve to drain the boiler.
3. Enter the boiler flue box and cut a hole in the side of the relevant smoke tube.
4. Clean the rim of the smoke tube with a wire brush.
5. Cut a circular plate (15 mm thick) of the same diameter as the smoke tube and chamfer the top edge to 30 degrees by grinding.
6. Fit the plate into the top of the smoke tube and weld it in position as shown.
7. Enter the boiler furnace and cut a similar hole in this end of the relevant smoke tube.
8. Repeat steps 4 to 6 for lower plate.
9. Refill boiler and check for leaks before start-up.
10. Start-up boiler and check for leaks when pressurized.

Note: Any temporary repair to smoke tubes or boiler tubes should receive more permanent attention as soon as conveniently possible.

REPAIRS IN WATER TUBE BOILER

Instruction for Plugging/ Repair of Water Tube Boiler & Economizer:

1. In case of tube failure, steam pressure has to be removed and the oil burner dismantled.
2. If the leakage is readily visible from the burner hole, the boiler can be emptied and repairs commence.
3. Otherwise, the boiler is given pressure by means of the feed pump. The position of the leakage will be indicated by the water flow.
4. This flow may not be visible from the burner hole. If it is not visible, remove the inspection door and enter the furnace. If the tube failure is still not found, then enter the generating tube section. From here the bottom of the membrane walls and generating tubes can be inspected for leakage.
5. If the leakage has resulted from the membrane walls or generating tube, the inspection door at the smoke connection pipe must be removed, and the generating tube/ membrane tube in which the failure has occurred is pointed out.
6. The leakage may also result from economizer.
7. By removing the inspection door at the bottom of the economizer, it can be determined which uptake has caused the leakage?
8. If necessary other inspection doors should be removed to point out the damage register.
9. When a damaged tube or convection register has been removed, and the remaining tube studs have been repaired/ plugged a new tube or register should be mounted as soon as possible.
10. Operation for longer periods with one or more registers missing involves the risk of further damage to the boiler due to increasing heat leads on the parts next to the ones removed.

Scope of Inspection of a Ship's Boiler

The boiler is one of the items of equipment on a ship which continuously keeps on running during sailing and in port. As it is running continuously, it has to be cleaned and inspected to check the condition of all internal working parts at regular intervals.

SCOPE OF INSPECTION

The scope of inspection is to clean the boiler's internal surfaces and to check for corrosion and scale formation in the boiler. As the boiler normally runs continuously, there are few chances to open the boiler. Thus, during the inspection all the important checks will be carried out and it will be made sure that the boiler will safely work without any problems until the next inspection. Routine inspection is important because salt formation and scaling inside the boiler tubes will reduce the heat transfer rate and ultimately damage the tubes due to overheating.

1. The inspection should include finding reasons for any abnormality found and should also ensure that any repair carried out does not affect that safe working order of the boiler.
2. A complete inspection means full internal and external examination of all parts of the boiler and accessories such as super-heaters, air heaters, and all mountings.
3. The examination may lead the inspector to require hydraulic testing of pressure parts or thickness gauging of plate or tubs that appear to be checked for good working condition.

The Inspection is not completed until the boiler has been examined under steam and the following items dealt with:

a. Pressure gauge checking against a test gauge.
b. Testing of water level indicators and protective devices.
c. Safety valves adjusted under steam to blow off at the required pressures.
d. The oil fuel burning system examined.
e. Testing of remote control gear for fuel shut off valves.

For a gas fired boiler, the chief engineer floats the safety valve at sea at the first opportunity. Survey record is not assigned until a statement is received from chief engineer about the pressure at which the safety valves were set.

INSPECTION CONSISTS OF:

a. Examination of the items.
b. Statement whether a problem/ defect exist.
c. Determining the cause of problem.
d. Define the repair and whether temporary/ permanent.

THE MAIN BENEFITS OF DOING INSPECTION:

By doing the inspection, we are manually cleaning the boiler scales and chemical cleaning of the salt formation in the boiler parts and making the boiler safe for operation. It also helps in checking the redundancy of the stand-by boiler. During the inspection the newly signed in crew members and the ship's engineer will also have a chance to see the internal parts of boiler.

1. Boiler must be sufficiently cleaned and dried to make a thorough examination possible.
2. Boiler should be manually wire-brushed to clean the internal surfaces.
3. In case of difficulty in manual cleaning, chemical cleaning with hydrochloric acid plus inhibitor to prevent acid attacking the metal without affecting removal of deposits is the best procedure.

4. For oil contamination, alkali boil-out using tri-sodium phosphate solution is essential prior to acid cleaning. Through water flushing must be carried out after acid cleaning to avoid acid concentration in crevices and captive spaces.
5. All internals that may interfere with the inspection have to be removed.
6. Wherever adequate visual examination is not possible, surveyor may have to resort to drilling, ultrasonic, or hydraulic testing.
7. All manhole doors and other doors must be opened for reasonable time previous to survey for ventilation.
8. If another boiler is under steam arrangement of locking bar and other security devices must be in position preventing the admission of steam or hot water to the boiler under survey. The smoke trunking, exhaust gas shut-off etc., must be in position and in proper working condition.
9. Plant's staff or repairer's staff should stand by the manhole in case of emergency and to take note for defects/ repairs required.

Before survey, the surveyor should acquaint himself with the boiler type in question (drawings carried on board) and during the survey it is advisable to follow a planned routine in order not to miss parts of the boiler or important items.

4

Pumps and Pumping Systems

PUMP

A pump is a device that moves fluids (liquids or gases), or sometimes slurries, by mechanical action. Pumps can be classified into three major groups according to the method they use to move the fluid: *direct lift*, *displacement*, and *gravity* pumps.

Pumps operate by some mechanism (typically reciprocating or rotary), and consume energy to perform mechanical work by moving the fluid. Pumps operate via many energy sources, including manual operation, electricity, engines, or wind power, come in many sizes, from microscopic for use in medical applications to large industrial pumps.

Mechanical pumps serve in a wide range of applications such as pumping water from wells, aquarium filtering, pond filtering andaeration, in the car industry for water-cooling and fuel injection, in the energy industry for pumping oil and natural gas or for operatingcooling towers. In the medical industry, pumps are used for biochemical processes in developing and manufacturing medicine, and as artificial replacements for body parts, in particular the artificial heart and penile prosthesis.

Single stage pump - When in a casing only one impeller is revolving then it is called single stage pump.

A small, electrically powered pump

A large, electrically driven pump (electropump) for waterworks near theHengsteysee, Germany

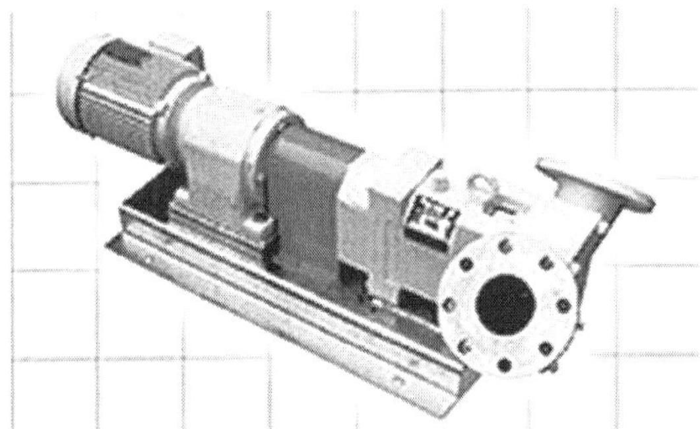

Horizontally mounted lobe pump (right) shown with its electric motor (left) and drive-shaft bearing (middle)

Double/ Multi stage pump - When in a casing two or more than two empeller is revolving than it is called double/ multi stage pump.

In biology, many different types of chemical and bio-mechanical pumps have evolved, and biomimicry is sometimes used in developing new types of mechanical pumps.

TYPES

Mechanical pumps may be submerged in the fluid they are pumping or be placed external to the fluid.

Pumps can be classified by their method of displacement into positive displacement pumps, impulse pumps, velocity pumps, gravity pumps, steam pumps and valveless pumps. There are two basic types of pumps: positive displacement and centrifugal. Although axial-flow pumps are frequently classified as a separate type, they have essentially the same operating principles as centrifugal pumps.

Positive displacement pump

A positive displacement pump makes a fluid move by trapping a fixed amount and forcing (displacing) that trapped volume into the discharge pipe.

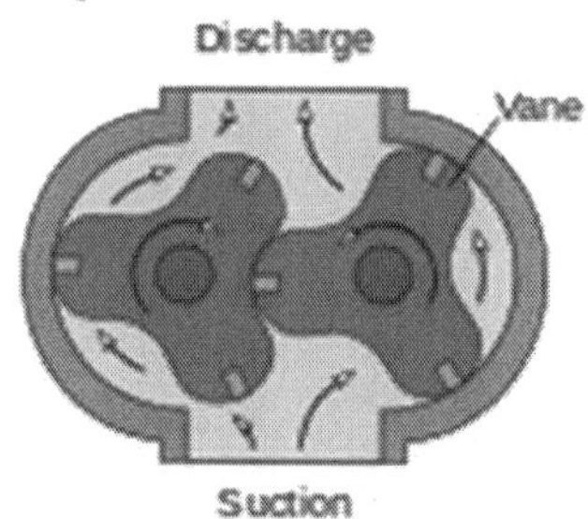

Fig. Lobe pump internals

Some positive displacement pumps use an expanding cavity on the suction side and a decreasing cavity on the discharge side. Liquid flows into the pump as the cavity on the suction side expands and the liquid flows out of the discharge as the cavity collapses. The volume is constant through each cycle of operation.

Positive displacement pump behavior and safety

Positive displacement pumps, unlike centrifugal or roto-dynamic pumps, theoretically can produce the same flow at a given speed (RPM) no matter what the discharge pressure. Thus, positive displacement pumps are *constant flow machines*. However, a slight increase in internal leakage as the pressure increases prevents a truly constant flow rate.

A positive displacement pump must not operate against a closed valve on the discharge side of the pump, because it has no shutoff head like centrifugal pumps. A positive displacement pump operating against a closed discharge valve continues to produce flow and the pressure in the discharge line increases until the line bursts, the pump is severely damaged, or both.

A relief or safety valve on the discharge side of the positive displacement pump is therefore necessary. The relief valve can be internal or external. The pump manufacturer normally has the option to supply internal relief or safety valves. The internal valve is usually only used as a safety precaution. An external relief valve in the discharge line, with a return line back to the suction line or supply tank provides increased safety.

Positive displacement types

A positive displacement pump can be further classified according to the mechanism used to move the fluid:

- *Rotary-type* positive displacement: internal gear, screw, shuttle block, flexible vane or sliding vane, circumferential piston, flexible impeller, helical twisted roots (e.g. the Wendelkolben pump) or liquid ring vacuum pumps
- *Reciprocating-type* positive displacement: piston or diaphragm pumps
- *Linear-type* positive displacement: rope pumps and chain pumps

Rotary positive displacement pumps

Rotary vane pump

These pumps move fluid using a rotating mechanism that creates a vacuum that captures and draws in the liquid.

Advantages: Rotary pumps are very efficient because they naturally remove air from the lines, eliminating the need to bleed the air from the lines manually.

Drawbacks: The nature of the pump requires very close clearances between the rotating pump and the outer edge, making it rotate at a slow, steady speed. If rotary pumps are operated at high speeds, the fluids cause erosion, which eventually causes enlarged clearances that liquid can pass through, which reduces efficiency.

Rotary positive displacement pumps fall into three main types:

- Gear pumps - a simple type of rotary pump where the liquid is pushed between two gears
- Screw pumps - the shape of the internals of this pump is usually two screws turning against each other to pump the liquid
- Rotary vane pumps - similar to scroll compressors, these have a cylindrical rotor encased in a similarly shaped housing. As the rotor orbits, the vanes trap fluid between the rotor and the casing, drawing the fluid through the pump.

Reciprocating positive displacement pumps

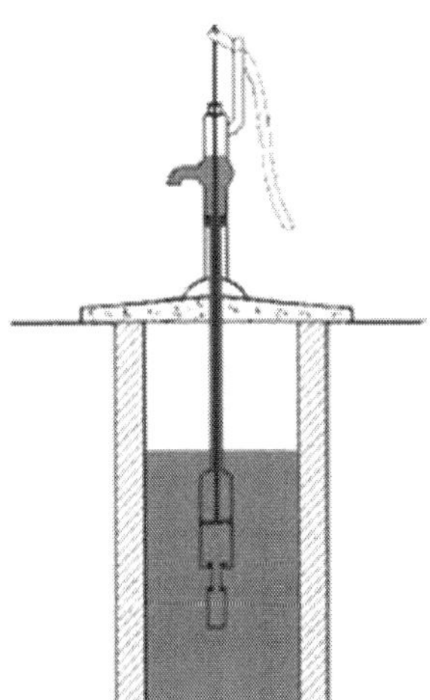

Simple hand pump

Old hand water pump (c. 1924) at the Colored School in Alapaha, Georgia, US Reciprocating pumps move the fluid using one or more oscillating pistons, plungers, or membranes (diaphragms), while valves restrict fluid motion to the desired direction.

Pumps in this category range from *simplex*, with one cylinder, to in some cases *quad* (four) cylinders, or more. Many reciprocating-type pumps are *duplex* (two) or *triplex* (three) cylinder. They can be either *single-acting* with suction during one direction of piston motion and discharge on the other, or *double-acting* with suction and discharge in both directions. The pumps can be powered manually, by air or steam, or by a belt driven by an engine. This type of pump was used extensively in the 19th century—in the early days of steam propulsion—as boiler feed water pumps. Now reciprocating pumps typically pump highly viscous fluids like concrete and heavy oils, and serve in special applications that demand low flow rates against high resistance. Reciprocating hand pumps were widely used to pump water from wells. Common bicycle pumps and foot pumps for inflation use reciprocating action.

These positive displacement pumps have an expanding cavity on the suction side and a decreasing cavity on the discharge side. Liquid flows into the pumps as the cavity on the suction side expands and the liquid flows out of the discharge as the cavity collapses. The volume is constant given each cycle of operation.

Typical reciprocating pumps are:

- *Plunger pumps* - a reciprocating plunger pushes the fluid through one or two open valves, closed by suction on the way back.
- *Diaphragm pumps* - similar to plunger pumps, where the plunger pressurizes hydraulic oil which is used to flex a diaphragm in the pumping cylinder. Diaphragm valves are used to pump hazardous and toxic fluids.
- *Piston pumps* displacement pumps - *usually simple devices for pumping small amounts of liquid or gel manually. The common hand soap dispenser is such a pump.*
- *Radial piston pumps*

Various positive displacement pumps

The positive displacement principle applies in these pumps:

- Rotary lobe pump
- Progressive cavity pump
- Rotary gear pump
- Piston pump
- Diaphragm pump
- Screw pump
- Gear pump
- Hydraulic pump
- Rotary vane pump
- Peristaltic pump
- Rope pump
- Flexible impeller pump

Gear pump

Gear pump

This is the simplest of rotary positive displacement pumps. It consists of two meshed gears that rotate in a closely fitted casing. The tooth spaces trap fluid and force it around the outer periphery. The fluid does not travel back on the meshed part, because the teeth mesh closely in the center. Gear pumps see wide use in car engine oil pumps and in various hydraulic power packs.

Screw pump

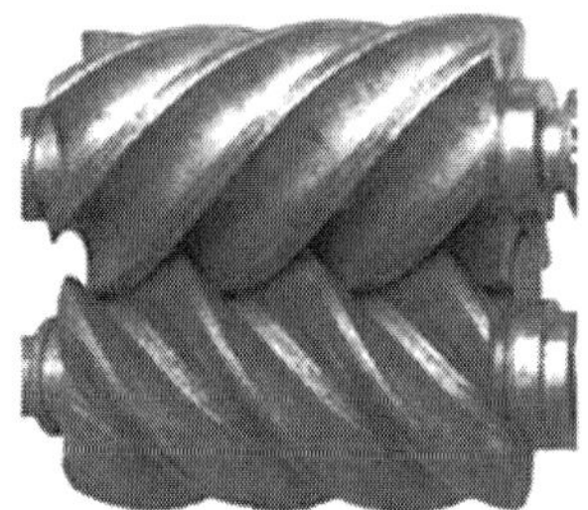

Screw pump

A screw pump is a more complicated type of rotary pump that uses two or three screws with opposing thread — e.g., one screw turns clockwise and the other counterclockwise. The screws are mounted on parallel shafts that have gears that mesh so the shafts turn together and everything stays in place. The screws turn on the shafts and drive fluid through the pump. As with other forms of rotary pumps, the clearance between moving parts and the pump's casing is minimal.

Progressing cavity pump

Widely used for pumping difficult materials, such as sewage sludge contaminated with large particles, this pump consists of a helical rotor, about ten times as long as its width. This can be visualized as a central core of diameter x with, typically, a curved spiral wound around of thickness half$x$, though in reality it is manufactured in single casting. This shaft fits inside a heavy duty rubber sleeve, of wall thickness also typically x. As the shaft rotates, the rotor gradually forces fluid up the rubber sleeve. Such pumps can develop very high pressure at low volumes.

Roots-type pumps

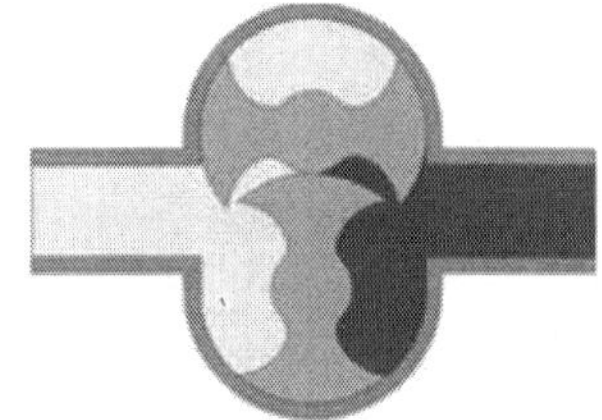

A Roots lobe pump

Named after the Roots brothers who invented it, this lobe pump displaces the liquid trapped between two long helical rotors, each fitted into the other when perpendicular at 90°, rotating inside a triangular shaped sealing line configuration, both at the point of suction and at the point of discharge. This design produces a continuous flow with equal volume and no vortex. It can work at low pulsation rates, and offers gentle performance that some applications require.

Applications include:

- High capacity industrial air compressors

- Roots superchargers on internal combustion engines.
- A brand of civil defense siren, the Federal Signal Corporation's Thunderbolt.

Peristaltic pump

360° Peristaltic Pump

A *peristaltic pump* is a type of positive displacement pump. It contains fluid within a flexible tube fitted inside a circular pump casing (though linear peristaltic pumps have been made). A number of *rollers*, *shoes*, or *wipers* attached to a rotor compresses the flexible tube. As the rotor turns, the part of the tube under compression closes (or *occludes*), forcing the fluid through the tube. Additionally, when the tube opens to its natural state after the passing of the cam it draws (*restitution*) fluid into the pump. This process is called peristalsis and is used in many biological systems such as the gastrointestinal tract.

Plunger pumps

Plunger pumps are reciprocating positive displacement pumps.

These consist of a cylinder with a reciprocating plunger. The suction and discharge valves are mounted in the head of the cylinder. In the suction stroke the plunger retracts and the suction valves open causing suction of fluid into the cylinder.

In the forward stroke the plunger pushes the liquid out of the discharge valve. Efficiency and common problems: With only one cylinder in plunger pumps, the fluid flow varies between maximum flow when the plunger moves through the middle positions, and zero flow when the plunger is at the end positions. A lot of energy is wasted when the fluid is accelerated in the piping system. Vibration and *water hammer* may be a serious problem. In general the problems are compensated for by using two or more cylinders not working in phase with each other.

Triplex-style plunger pumps

Triplex plunger pumps use three plungers, which reduces the pulsation of single reciprocating plunger pumps. Adding a pulsation dampener on the pump outlet can further smooth the *pump ripple*, or ripple graph of a pump transducer. The dynamic relationship of the high-pressure fluid and plunger generally requires high-quality plunger seals. Plunger pumps with a larger number of plungers have the benefit of increased flow, or smoother flow without a pulsation dampener. The increase in moving parts and crankshaft load is one drawback.

Car washes often use these triplex-style plunger pumps (perhaps without pulsation dampeners). In 1968, William Bruggeman significantly reduced the size of the triplex pump and increased the lifespan so that car washes could use equipment with smaller footprints. Durable high pressure seals, low pressure seals and oil seals, hardened crankshafts, hardened connecting rods, thick ceramic plungers and heavier duty ball and roller bearings improve reliability in triplex pumps. Triplex pumps now are in a myriad of markets across the world.

Triplex pumps with shorter lifetimes are commonplace to the home user. A person who uses a home pressure washer for 10 hours a year may be satisfied with a pump that lasts 100 hours between rebuilds. Industrial-grade or continuous duty triplex pumps on the other end of the quality spectrum may run for as much as 2,080 hours a year.

The oil and gas drilling industry uses massive semi trailer-transported triplex pumps called mud pumps to pump drilling mud, which cools the drill bit and carries the cuttings back to the surface. Drillers use triplex or even quintuplex pumps to inject water and solvents deep into shale in the extraction process called *fracking*.

Compressed-air-powered double-diaphragm pumps

One modern application of positive displacement diaphragm pumps is compressed-air-powered double-diaphragm pumps. Run on compressed air these pumps are intrinsically safe by design, although all manufacturers offer ATEX certified models to comply with industry regulation. These pumps are relatively inexpensive and can perform a wide variety of duties, from pumping water out of bunds, to pumping hydrochloric acid from secure storage (dependent on how the pump is manufactured – elastomers/ body construction). Lift is normally limited to roughly 6m although heads can reach almost 200 psi (1.4 MPa).

Rope pumps

Devised in China as chain pumps over 1000 years ago, these pumps can be made from very simple materials: A rope, a wheel and a PVC pipe are sufficient to make a simple rope pump. For this reason they have become extremely popular around the world since the 1980s. Rope pump efficiency has been studied by grass roots organizations and the techniques for making and running them have been continuously improved.

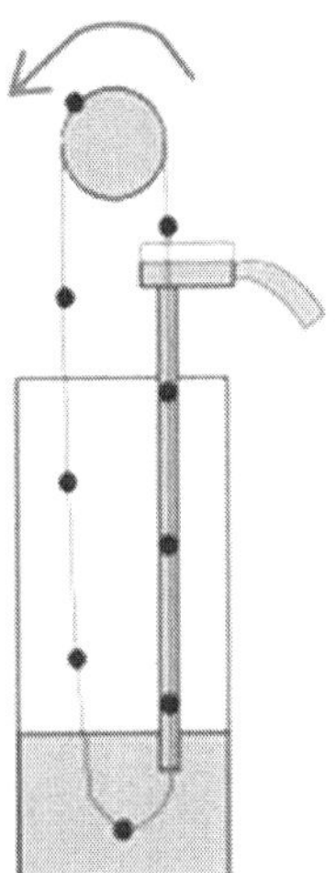

Rope pump schematic

Impulse pumps

Impulse pumps use pressure created by gas (usually air). In some impulse pumps the gas trapped in the liquid (usually water), is released and accumulated somewhere in the pump, creating a pressure that can push part of the liquid upwards.

Conventional impulse pumps include:

- *Hydraulic ram pumps* – kinetic energy of a low-head water supply is stored temporarily in an air-bubble hydraulic accumulator, then used to drive water to a higher head.
- *Pulser pumps* – run with natural resources, by kinetic energy only.
- *Airlift pumps* – run on air inserted into pipe, which pushes the water up when bubbles move upward

Instead of a gas accumulation and releasing cycle, the pressure can be created by burning of hydrocarbons. Such combustion driven pumps directly transmit the impulse form a combustion event through the actuation membrane to the pump fluid. In order to allow this direct transmission, the pump needs to be almost entirely made of an elastomer (e.g. silicone rubber). Hence, the combustion causes the membrane to expand and thereby pumps the fluid out of the adjacent pumping chamber. The first combustion-driven soft pump was developed by ETH Zurich.

Hydraulic ram pumps

A hydraulic ram is a water pump powered by hydropower.

It takes in water at relatively low pressure and high flow-rate and outputs water at a higher hydraulic-head and lower flow-rate. The device uses the water hammer effect to develop pressure that lifts a portion of the input water that powers the pump to a point higher than where the water started. The hydraulic ram is sometimes used in remote areas, where there is both a source of low-head hydropower, and a need for pumping water to a destination higher in elevation than the source. In this situation, the ram is often useful, since it requires no outside source of power other than the kinetic energy of flowing water.

VELOCITY PUMPS

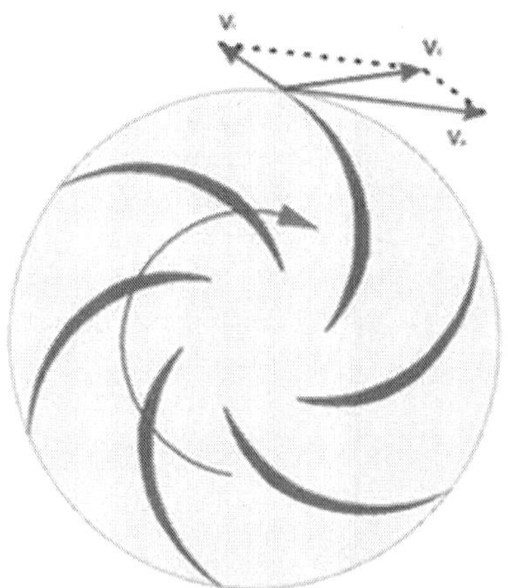

A centrifugal pump uses an impellerwith backward-swept arms

Rotodynamic pumps (or dynamic pumps) are a type of velocity pump in which kinetic energy is added to the fluid by increasing the flow velocity. This increase in energy is converted to a gain in potential energy (pressure) when the velocity is reduced prior to or as the flow exits the pump into the discharge pipe. This conversion of kinetic energy to pressure is explained by the *First law of thermodynamics*, or more specifically by *Bernoulli's principle*.

Dynamic pumps can be further subdivided according to the means in which the velocity gain is achieved.

These types of pumps have a number of characteristics:

1. Continuous energy
2. Conversion of added energy to increase in kinetic energy (increase in velocity)
3. Conversion of increased velocity (kinetic energy) to an increase in pressure head

A practical difference between dynamic and positive displacement pumps is how they operate under closed valve conditions. Positive displacement pumps physically displace fluid, so closing a valve downstream of a positive displacement pump produces a continual pressure build up that can cause mechanical failure of pipeline or pump. Dynamic pumps differ in that they can be safely operated under closed valve conditions (for short periods of time).

Radial-flow pumps

These are also referred to as *centripetal design* pumps. The fluid enters along the axis or center, is accelerated by the impeller and exits at right angles to the shaft(radially). Radial-flow pumps operate at higher pressures and lower flow rates than axial- and mixed-flow pumps.

Axial-flow pumps

These are also referred to as All fluid pumps The fluid is pushed outward or inward and move fluid axially. They operate at much lower pressures and higher flow rates than radial-flow (centripetal) pumps.

Mixed-flow pumps

Mixed-flow pumps function as a compromise between radial and axial-flow pumps. The fluid experiences both radial acceleration and lift and exits the impeller somewhere between 0 and 90 degrees from the axial direction. As a consequence mixed-flow pumps operate at higher pressures than axial-flow pumps while delivering higher discharges than radial-flow pumps. The exit angle of the flow dictates the pressure head-discharge characteristic in relation to radial and mixed-flow.

Eductor-jet pump

This uses a jet, often of steam, to create a low pressure. This low pressure sucks in fluid and propels it into a higher pressure region.

Gravity pumps

Gravity pumps include the *syphon* and *Heron's fountain*. The *hydraulic ram* is also sometimes called a gravity pump; in a gravity pump the water is lifted by gravitational force.

Steam pumps

Steam pumps have been for a long time mainly of historical interest. They include any type of pump powered by a steam engine and also pistonless pumps such as Thomas Savery's or the Pulsometer steam pump. Recently there has been a resurgence of interest in low power solar steam pumps for use in smallholder irrigation in developing countries. Previously small steam engines have not been viable because of escalating inefficiencies as vapour engines decrease in size. However the use of modern engineering materials coupled with alternative engine configurations has meant that these types of system are now a cost effective opportunity.

Valveless pumps

Valveless pumping assists in fluid transport in various biomedical and engineering systems. In a valveless pumping system, no valves (or physical occlusions) are present to regulate the flow direction. The fluid pumping efficiency of a valveless system, however, is not necessarily lower than that having valves. In fact, many fluid-dynamical systems in nature and engineering more or less rely upon valveless pumping to transport the working fluids therein. For instance, blood circulation in the cardiovascular system is maintained to some extent even when the heart's valves fail. Meanwhile, the embryonic vertebrate heart begins pumping blood long before the development of discernible chambers and valves. In microfluidics, valveless impedance pumps have been fabricated, and are expected to be particularly suitable for handling sensitive biofluids. Ink jet printers operating on the Piezoelectric transducer principle also use valveless pumping. The pump chamber is emptied through the printing jet due to reduced flow impedance in that direction and refilled by capillary action..

PUMP REPAIRS

Examining pump repair records and mean time between failures (MTBF) is of great importance to responsible and conscientious pump users. In view of that fact, the preface to the 2006 Pump User's Handbook alludes to "pump failure" statistics. For the sake of convenience, these failure statistics often are translated into MTBF (in this case, installed life before failure).

In early 2005, Gordon Buck, John Crane Inc.'s chief engineer for Field Operations in Baton Rouge, LA, examined the repair records for a number of refinery and chemical plants to obtain meaningful reliability data for centrifugal pumps. A total of 15 operating plants having nearly 15,000 pumps were included in the survey. The smallest of these plants had about 100 pumps; several plants had over 2000. All facilities were located in the United States. In addition, considered as "new", others as "renewed" and still others as "established". Many of these plants—but not all—had an alliance arrangement with John Crane. In some cases, the alliance contract included having a John Crane Inc. technician or engineer on-site to coordinate various aspects of the program.

Not all plants are refineries, however, and different results occur elsewhere. In chemical plants, pumps have traditionally been "throw-away" items as chemical attack limits life. Things have improved in recent years, but the somewhat restricted space available in "old" DIN and ASME-standardized stuffing boxes places limits on the type of seal that fits. Unless the pump user upgrades the seal chamber, the pump only accommodates more compact and simple versions. Without this upgrading, lifetimes in chemical installations are generally around 50 to 60 percent of the refinery values.

Unscheduled maintenance is often one of the most significant costs of ownership, and failures of mechanical seals and bearings are among the major causes. Keep in mind the potential value of selecting pumps that cost more initially, but last much longer between repairs. The MTBF of a better pump may be one to four years longer than that of its non-upgraded counterpart. Consider that published average values of avoided pump failures range from US$2600 to US$12,000. This does not include lost opportunity costs. One pump fire occurs per 1000 failures. Having fewer pump failures means having fewer destructive pump fires.

As has been noted, a typical pump failure based on actual year 2002 reports, costs US$5,000 on average. This includes costs for material, parts, labor and overhead. Extending a pump's MTBF from 12 to 18 months would save US$1,667 per year — which might be greater than the cost to upgrade the centrifugal pump's reliability.

APPLICATIONS

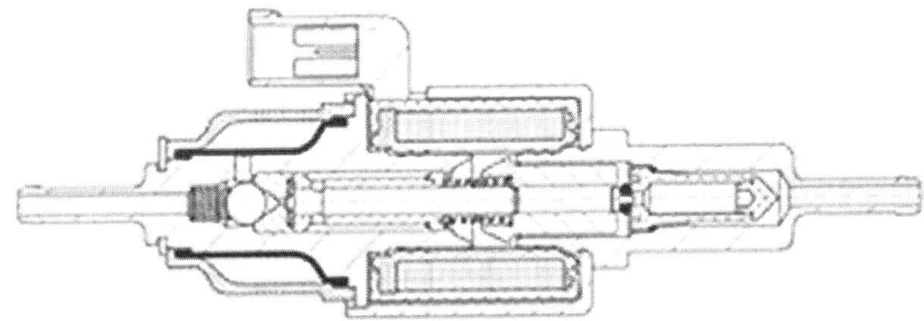

Metering pump for gasoline andadditives.

Pumps are used throughout society for a variety of purposes. Early applications includes the use of the windmill or watermill to pump water. Today, the pump is used for irrigation, water supply, gasoline supply, air conditioning systems, refrigeration (usually called a compressor), chemical movement, sewage movement, flood control, marine services, etc.

Because of the wide variety of applications, pumps have a plethora of shapes and sizes: from very large to very small, from handling gas to handling liquid, from high pressure to low pressure, and from high volume to low volume.

Priming a pump

Typically, a liquid pump can't simply draw air. The feed line of the pump and the internal body surrounding the pumping mechanism must first be filled with the liquid that requires pumping: An operator must introduce liquid into the system to initiate the pumping. This is called *priming* the pump. Loss of prime is usually due to ingestion of air into the pump. The clearances and displacement ratios in pumps for liquids, whether thin or more viscous, usually cannot displace air due to its compressibility. This is the case with most velocity (rotodynamic) pumps — for example, centrifugal pumps.

Positive–displacement pumps, however, tend to have sufficiently tight sealing between the moving parts and the casing or housing of the pump that they can be described as*self-priming*. Such pumps can also serve as *priming* *pumps*, so called when they are used to fulfill that need for other pumps in lieu of action taken by a human operator.

Pumps as public water supplies

One sort of pump once common worldwide was a hand-powered water pump, or 'pitcher pump'. It was commonly installed over community water wells in the days before piped water supplies.

In parts of the British Isles, it was often called *the parish pump*. Though such community pumps are no longer common, people still used the expression *parish pump* to describe a place or forum where matters of local interest are discussed.

Because water from pitcher pumps is drawn directly from the soil, it is more prone to contamination. If such water is not filtered and purified, consumption of it might lead to gastrointestinal or other water-borne diseases. A notorious case is the 1854 Broad Street cholera outbreak. At the time it was not known how cholera was transmitted, but physician John Snow suspected contaminated water and had the handle of the public pump he suspected removed; the outbreak then subsided.

Modern hand-operated community pumps are considered the most sustainable low-cost option for safe water supply in resource-poor settings, often in rural areas in developing countries. A hand pump opens access to deeper groundwater that is often not polluted and also improves the safety of a well by protecting the water source from contaminated buckets. Pumps such as the Afridev pump are designed to be cheap to build and install, and easy to maintain with simple parts. However, scarcity of spare parts for these type of pumps in some regions of Africa has diminished their utility for these areas.

Sealing multiphase pumping applications

Multiphase pumping applications, also referred to as tri-phase, have grown due to increased oil drilling activity. In addition, the economics of multiphase production is attractive to upstream operations as it leads to simpler, smaller in-field installations, reduced equipment costs and improved production rates. In essence, the multiphase pump can accommodate all fluid stream properties with one piece of equipment, which has a smaller footprint. Often, two smaller multiphase pumps are installed in series rather than having just one massive pump.

For midstream and upstream operations, multiphase pumps can be located onshore or offshore and can be connected to single or multiple wellheads. Basically, multiphase pumps are used to transport the untreated flow stream produced from oil wells to downstream processes or gathering facilities. This means that the pump may handle a flow stream (well stream) from 100 percent gas to 100 percent liquid and every imaginable combination in between. The flow stream can also contain abrasives such as sand and dirt. Multiphase pumps are designed to operate under changing/fluctuating process conditions. Multiphase pumping also helps eliminate emissions of greenhouse gases as operators strive to minimize the flaring of gas and the venting of tanks where possible.

Types and features of multiphase pumps

Helico-Axial Pumps (Centrifugal) A rotodynamic pump with one single shaft that requires two mechanical seals, this pump uses an open-type axial impeller. It's often called a*Poseidon pump*, and can be described as a cross between an axial compressor and a centrifugal pump.

Twin Screw (Positive Displacement) The twin screw pump is constructed of two inter-meshing screws that move the pumped fluid. Twin screw pumps are often used when pumping conditions contain high gas volume fractions and fluctuating inlet conditions. Four mechanical seals are required to seal the two shafts.

Progressive Cavity Pumps (Positive Displacement) Progressive cavity pumps are single-screw types typically used in shallow wells or at the surface. This pump is mainly used on surface applications where the pumped fluid may contain a considerable amount of solids such as sand and dirt.

Electric Submersible Pumps (Centrifugal) These pumps are basically multistage centrifugal pumps and are widely used in oil well applications as a method for artificial lift. These pumps are usually specified when the pumped fluid is mainly liquid.

Buffer Tank A buffer tank is often installed upstream of the pump suction nozzle in case of a slug flow. The buffer tank breaks the energy of the liquid slug, smooths any fluctuations in the incoming flow and acts as a sand trap.

As the name indicates, multiphase pumps and their mechanical seals can encounter a large variation in service conditions such as changing process fluid composition, temperature variations, high and low operating pressures and exposure to abrasive/erosive media. The challenge is selecting the appropriate mechanical seal arrangement and support system to ensure maximized seal life and its overall effectiveness.

SPECIFICATIONS

Pumps are commonly rated by horsepower, flow rate, outlet pressure in metres (or feet) of head, inlet suction in suction feet (or metres) of head. The head can be simplified as the number of feet or metres the pump can raise or lower a column of water at atmospheric pressure.

From an initial design point of view, engineers often use a quantity termed the specific speed to identify the most suitable pump type for a particular combination of flow rate and head.

PUMPING POWER

The power imparted into a fluid increases the energy of the fluid per unit volume. Thus the power relationship is between the conversion of the mechanical energy of the pump mechanism and the fluid elements within the pump. In general, this is governed by a series of simultaneous differential equations, known as the Navier–Stokes equations. However a more simple equation relating only the different energies in the fluid, known as Bernoulli's equation can be used. Hence the power, P, required by the pump:

$$P = \frac{\Delta p Q}{\eta}$$

where Δp is the change in total pressure between the inlet and outlet (in Pa), and Q, the volume flow-rate of the fluid is given in m^3/s. The total pressure may have gravitational,static pressure and kinetic energy components; i.e. energy is distributed between change in the fluid's gravitational potential energy (going up or down hill), change in velocity, or change in static pressure. ç is the pump efficiency, and may be given by the manufacturer's information, such as in the form of a pump curve, and is typically derived from either fluid dynamics simulation (i.e. solutions to the Navier–Stokes for the particular pump geometry), or by testing. The efficiency of the pump depends upon the pump's configuration and operating conditions (such as rotational speed, fluid density and viscosity etc.)

$$\Delta P = \frac{(v_2^2 - v_1^2)}{2} + \Delta z g + \frac{\Delta p_{\text{static}}}{\rho}$$

For a typical "pumping" configuration, the work is imparted on the fluid, and is thus positive. For the fluid imparting the work on the pump (i.e. a turbine), the work is negative. Power required to drive the pump is determined by dividing the output power by the pump efficiency. Furthermore, this definition encompasses pumps with no moving parts, such as a siphon.

PUMP EFFICIENCY

Pump efficiency is defined as the ratio of the power imparted on the fluid by the pump in relation to the power supplied to drive the pump. Its value is not fixed for a given pump, efficiency is a function of the discharge and therefore also operating head. For centrifugal pumps, the efficiency tends to increase with flow rate up to a point midway through the operating range (peak efficiency) and then declines as flow rates rise further. Pump performance data such as this is usually supplied by the manufacturer before pump selection. Pump efficiencies tend to decline over time due to wear (e.g. increasing clearances as impellers reduce in size). When a system design includes a centrifugal pump, an important issue it its design is matching the *head loss-flow characteristic* with the pump so that it operates at or close to the point of its maximum efficiency. Pump efficiency is an important aspect and pumps should be regularly tested. Thermodynamic pump testing is one method.

DIFFERENT TYPES OF PUMP SYSTEMS

There are many types of centrifugal pump systems. A typical industrial pump system. There are many variations on this including all kinds of equipment that can be hooked up to these systems that are not shown. A pump after all is only a single component of a process although an important and vital one. The pumps' role is to provide sufficient pressure to move the fluid through the system at the desired flow rate.

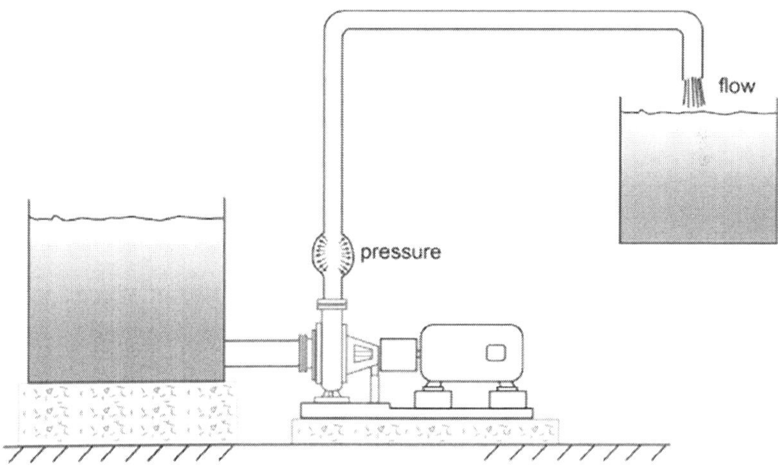

Back in the old days domestic water supply was simpler...aaah the good old days. Goodnight John boy..

Domestic water systems take their water from various sources at different levels depending on the water table and terrain contours.

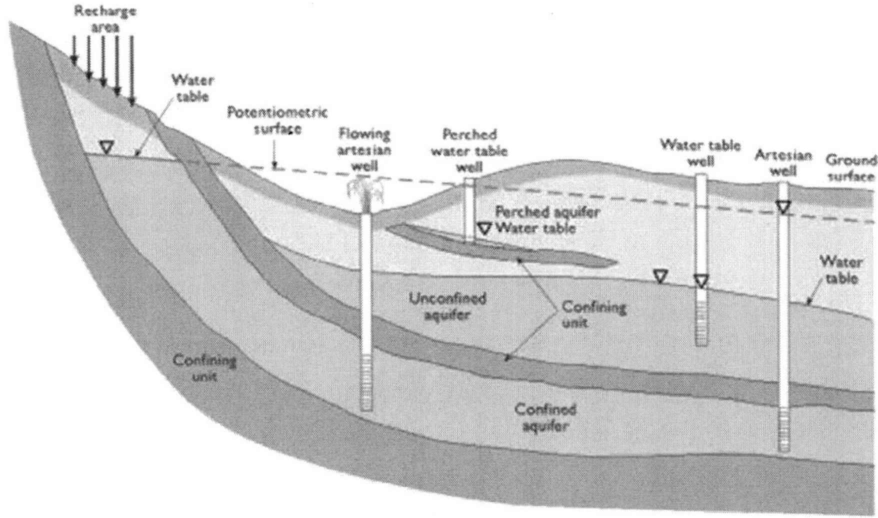

The system is a typical domestic water supply system that takes it's water from a shallow well (25 feet down max.) using an end suction centrifugal pump. A jet pump works well in this application.

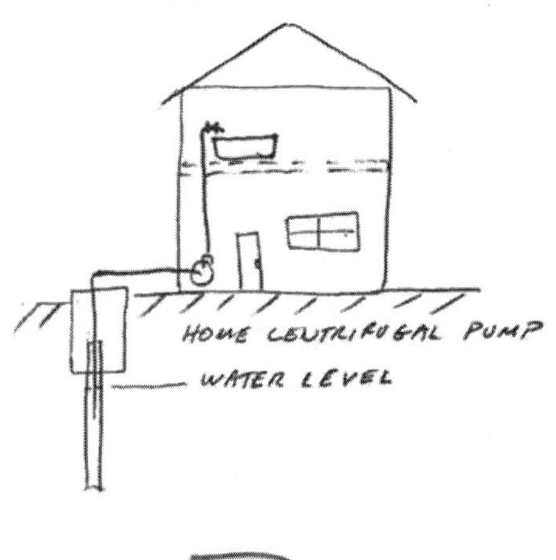

Fig. Typical jet pump.

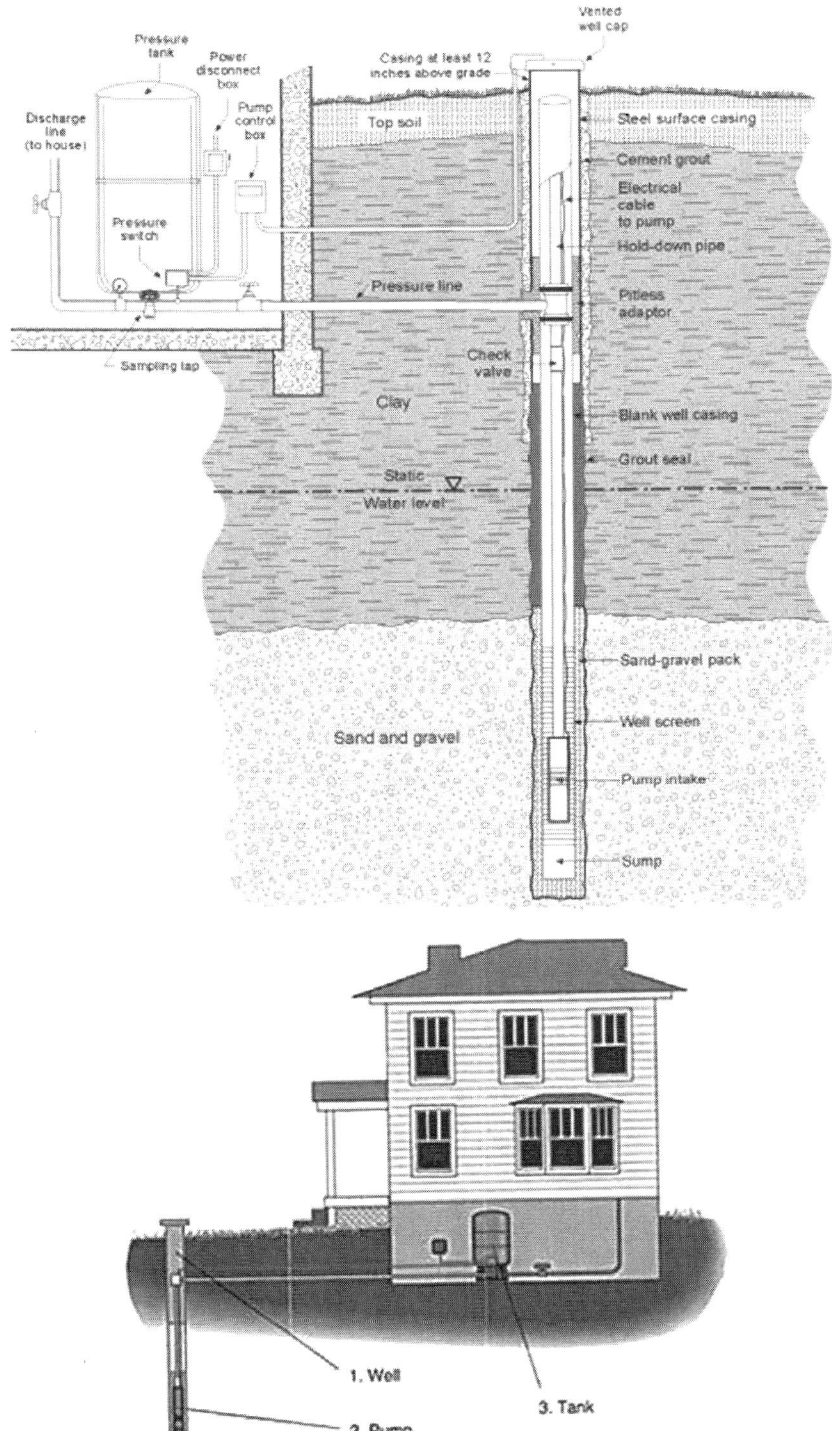
Vented well cap
Pressure tank
Power disconnect box
Casing at least 12 inches above grade
Pump control box
Discharge line (to house)
Top soil
Steel surface casing
Cement grout
Electrical cable to pump
Pressure switch
Hold-down pipe
Pressure line
Pitless adaptor
Sampling tap
Check valve
Clay
Blank well casing
Grout seal
Static
Water level
Sand-gravel pack
Well screen
Sand and gravel
Pump intake
Sump
1. Well
3. Tank
2. Pump

Fig. Typical deep well submersible pump

Pressure, friction and flow

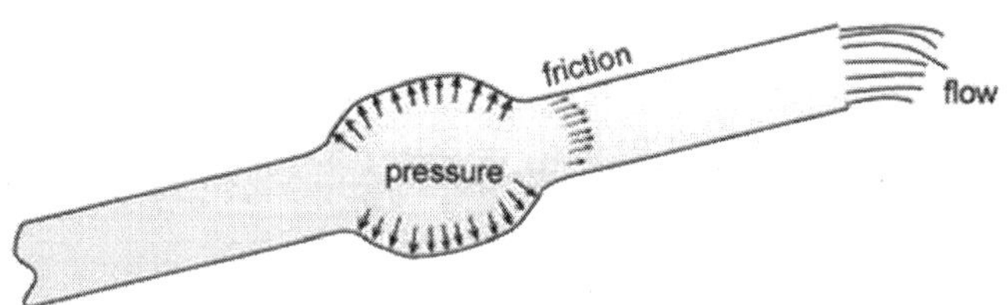

Pressure, friction and flow are three important characteristics of a pump system. Pressure is the driving force responsible for the movement of the fluid. Friction is the force that slows down fluid particles. Flow rate is the amount of volume that is displaced per unit time. The unit of flow in North America, at least in the pump industry, is the US gallon per minute, USgpm. From now on I will just use gallons per minute or gpm. In the metric system, flow is in liters per second (L/s) or meters cube per hour (m^3/h).

Pressure is often expressed in pounds per square inch (psi) in the Imperial system and kiloPascals (kPa) in the metric system. In the Imperial system of measurement, the unit psig or pounds per square inch gauge is used, it means that the pressure measurement is relative to the local atmospheric pressure, so that 5 psig is 5 psi above the local atmospheric pressure. In the metric system, the kPa unit scale is a scale of absolute pressure measurement and there is no kPag, but many people use the kPa as a relative measurement to the local atmosphere and don't bother to specify this. This is not a fault of the metric system but the way people use it. The term pressure loss or pressure drop is often used, this refers to the decrease in pressure in the system due to friction. In a pipe or tube that is at the same level, your garden hose for example, the pressure is high at the tap and zero at the hose outlet, this decrease in pressure is due to friction and is the pressure loss. As an example of the use of pressure and flow units, the pressure available to domestic water systems varies greatly depending on your location with respect to the water treatment plant. It can vary between 30 and 70 psi or more. The following table gives the expected flow rate that you would obtain for different pipe sizes assuming the pipe or tube is kept at the same level as the connection to the main water pressure supply and has a 100 feet of length.

Flow rate based on available pressure and pipe size

Nom. dia. (in)	Inside dia. (in)	Flow rate (gpm) @ 30 psi	Flow rate (gpm) @ 40 psi	Flow rate (gpm) @ 50 psi	Flow rate (gpm) @ 60 psi	Flow rate (gpm) @ 70 psi
1/4	0.311	1.3	1.5	1.7	1.9	2
1/2	0.527	5.4	6.3	7	7.7	8.5
3/4	0.745	13.6	15.6	17.7	20	21
1	0.995	29	34	38	42	46
1 1/2	1.6	103	119	135	147	160

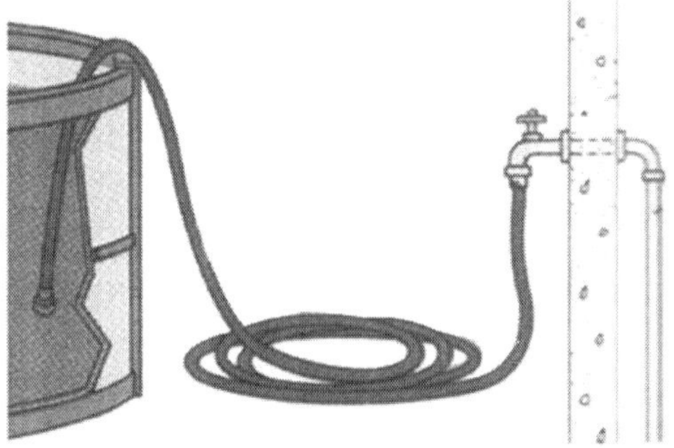

The unit of friction is....Sorry, I think I need to wait 'til we get closer to the end to explain the reasoning behind this unit.

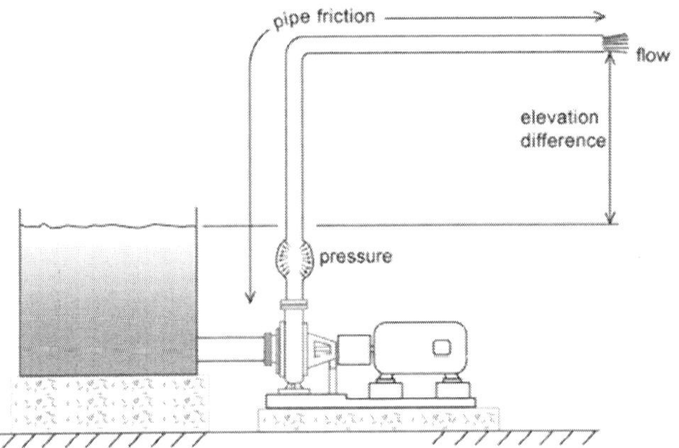

Pressure provides the driving force to overcome friction and elevation difference. It's responsible for driving the fluid through the system, the pump provides the pressure. Pressure is increased when fluid particles are forced closer together. For example, in a fire extinguisher work or energy has been spent to pressurize the liquid chemical within, that energy can be stored and used later. Is it possible to pressurize a liquid within a container that is open? Yes. A good example is a syringe, as you push down on the plunger the pressure increases, and the harder you have to push. There is enough friction as the fluid moves through the needle to produce a great deal of pressure in the body of the syringe

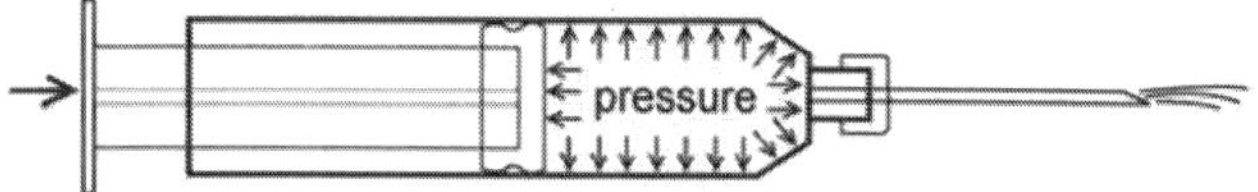

If we apply this idea to the pump system, even though the discharge pipe end is open, it is possible to have pressure at the pump discharge because there is sufficient friction in the system and elevation difference.

WHAT IS FRICTION IN A PUMP SYSTEM

Friction is always present, even in fluids, it is the force that resists the movement of objects.

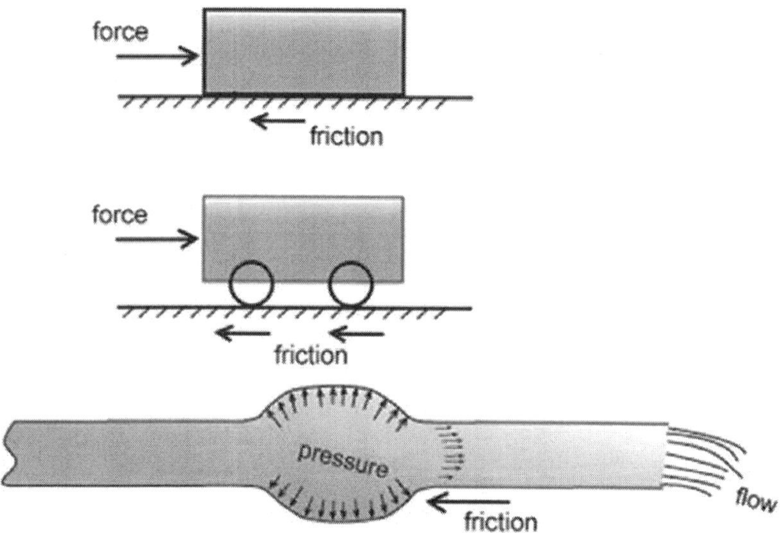

When you move a solid on a hard surface, there is friction between the object and the surface. If you put wheels on it, there will be less friction. In the case of moving fluids such as water, there is even less friction but it can become significant for long pipes. Friction can also be high for short pipes which have a high flow rate and small diameter as in the syringe example.

In fluids, friction occurs between fluid layers that are traveling at different velocities within the pipe. There is a natural tendency for the fluid velocity to be higher in the center of the pipe than near the wall of the pipe. Friction will also be high for viscous fluids and fluids with suspended particles.

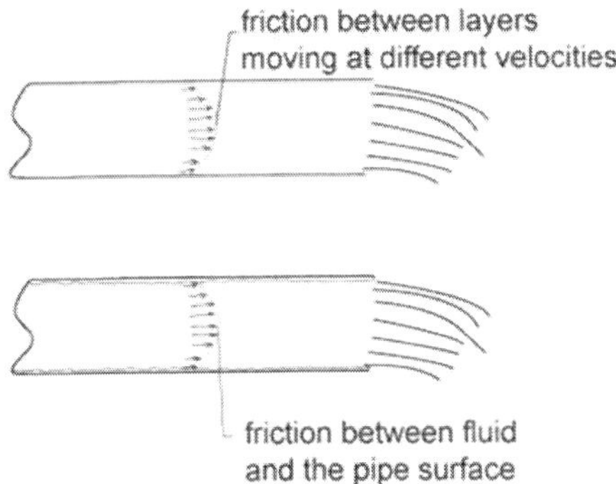

Another cause of friction is the interaction of the fluid with the pipe wall, the rougher the pipe, the higher the friction.

WHAT IS FRICTION IN A PUMP SYSTEM

Another cause of friction is all the fittings (elbows, tees, y's, etc) required to get the fluid from point A to B. Each one has a particular effect on the fluid streamlines. For example in the case of the elbow, the fluid particles that are closest to the tight inner radius of the elbow lift off from the pipe surface forming small vortices that consume energy. This energy loss is small for one elbow but if you have several elbows and other fittings the total can become significant. Generally speaking they rarely represent more then 30% of the total friction due to the overall pipe length.

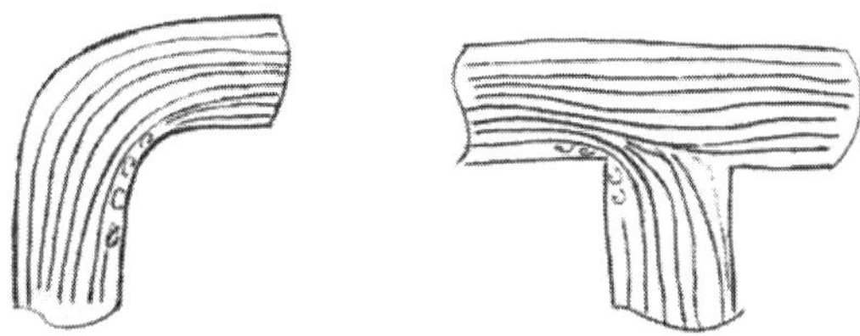

ENERGY AND HEAD IN PUMP SYSTEMS

Energy and head are two terms that are often used in pump systems. We use energy to describe the movement of liquids in pump systems because it is easier than any other method. There are four forms of energy in pump systems: pressure, elevation, friction and velocity.

Pressure is produced at the bottom of the reservoir because the liquid fills up the container completely and its weight produces a force that is distributed over a surface which is pressure. This type of pressure is called static pressure. Pressure energy is the energy that builds up when liquid or gas particles are moved slightly closer to each other and as a result they push outwards in their environment. A good example is a fire extinguisher, work was done to get the liquid into the container and then to pressurize it. Once the container is closed the pressure energy is available for later use.

Elevation energy is the energy that is available to a liquid when it is at a certain height. If you let it discharge it can drive something useful like a turbine producing electricity.

Friction energy is the energy that is lost to the environment due to the movement of the liquid through pipes and fittings in the system.

Velocity energy is the energy that moving objects have. When a baseball is thrown by a pitcher he gives it velocity energy also called kinetic energy. When water comes out of a garden hose, it has velocity energy.

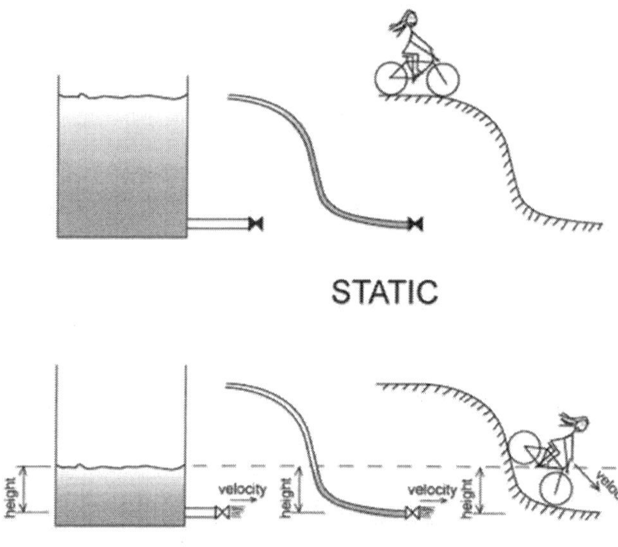

We see a tank full of water, a tube full of water and a cyclist at the top of a hill. The tank produces pressure at the bottom and so does the tube. The cyclist has elevation energy which he will be using as soon as he moves.

As we open the valve at the tank bottom the fluid leaves the tank with a certain velocity, in this case pressure energy is converted to velocity energy. The same thing happens with the tube. In the case of the cyclist, the elevation energy is gradually converted to velocity energy.

The three forms of energy: elevation, pressure and velocity interact with each other in liquids. For solid objects there is no pressure energy because they don't extend outwards like liquids filling up all the available space and therefore they are not subject to the same kind of pressure changes.

The energy that the pump must supply is the friction energy plus the elevation energy.

PUMP ENERGY = FRICTION ENERGY + ELEVATION ENERGY

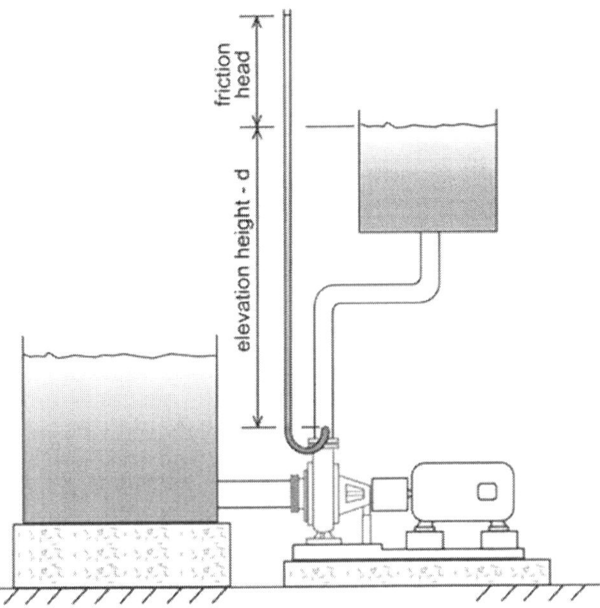

You are probably thinking where is the velocity energy in all this. Well if the liquid comes out of the system at high velocity then we would have to consider it but this is not a typical situation and we can neglect this for the systems.

The last word on this topic, it is actually the velocity energy difference that we would need to consider. The velocities at point 1 and point 2 are the result of the position of the fluid particles at points 1 and 2 and the action of the pump. The difference between these two velocity energies is an energy deficiency that the pump must supply but as you can see the velocities of these two points will be quite small.

Now what about head? Head is actually a way to simplify the use of energy. To use energy we need to know the weight of the object displaced.

Elevation energy E.E. is the weight of the object W times the distance d:

$$EE = W \times d$$

Friction energy FE is the force of friction F times the distance the liquid is displaced or the pipe length l:

$$FE = F \times l$$

Head is defined as energy divided by weight or the amount of energy used to displace a object divided by its weight. For elevation energy, the elevation head EH is:

$$EH = W \times d / W = d$$

For friction energy, the friction head FH is the friction energy divided by the weight of liquid displaced:

$$FH = FE/W = F \times l/ W$$

The friction force F is in pounds and W the weight is also in pounds so that the unit of friction head is feet. This represents the amount of energy that the pump has to provide to overcome friction.

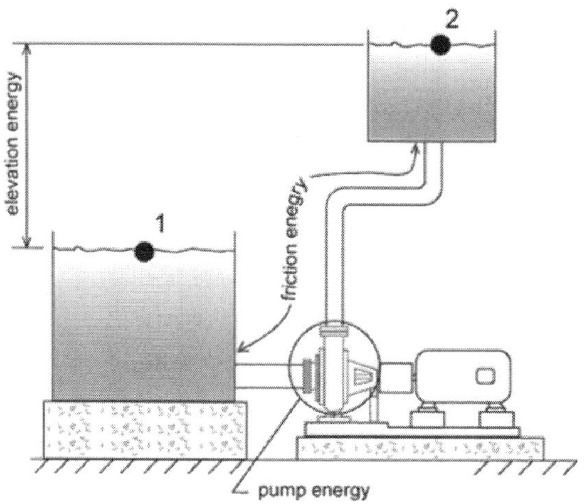

I know you are thinking this doesn't make sense, how can feet represent energy? If I attach a tube to the discharge side of a pump, the liquid will rise in the tube to a height that exactly balances the pressure at the pump discharge. Part of the height of liquid in the tube is due to the elevation height required (elevation head) and the other is the friction head and as you can see both are expressed in feet and this is how you can measure them.

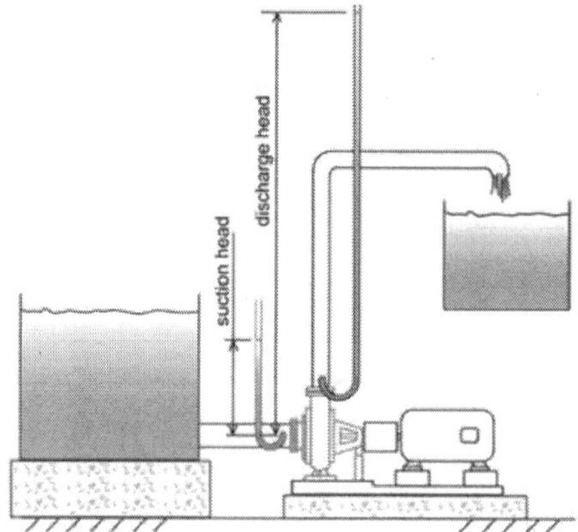

STATIC HEAD

Webster's dictionary definition of head is: "a body of water kept in reserve at a height".

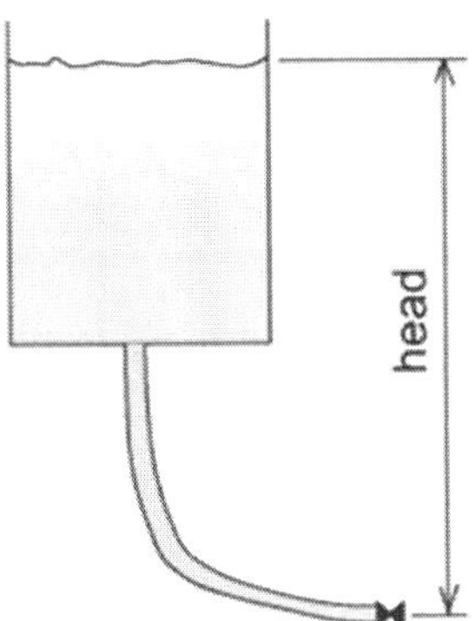

It is expressed in terms of feet in the Imperial system and meters in the metric system. Because of its height and weight the fluid produces pressure at the low point. The higher the reservoir, the higher the pressure.

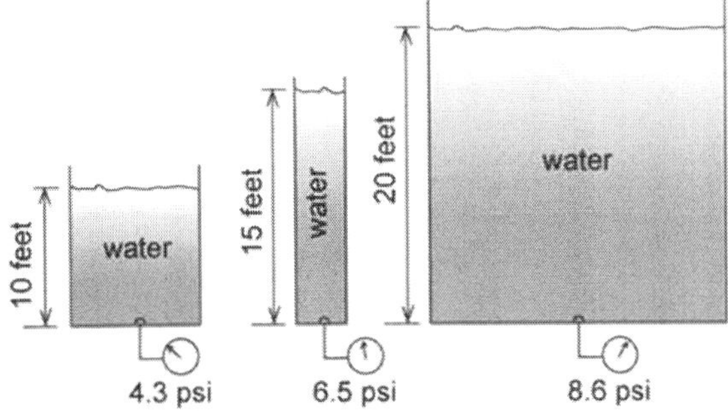

The amount of pressure at the bottom of a reservoir is independent of its shape, for the same liquid level, the pressure at the bottom will be the same. This is important since in complex piping systems it will always be possible to know the pressure at the bottom if we know the height.

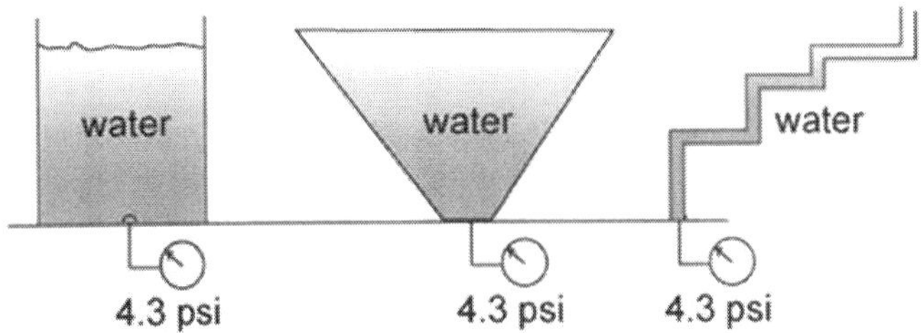

When a pump is used to displace a liquid to a higher level it is usually located at the low point or close to it. The head of the reservoir which is called static head will produce pressure on the pump that will have to be overcome once the pump is started.

To distinguish between the pressure energy produced by the discharge tank and suction tank, the head on the discharge side is called the discharge static head and on the suction side the suction static head.

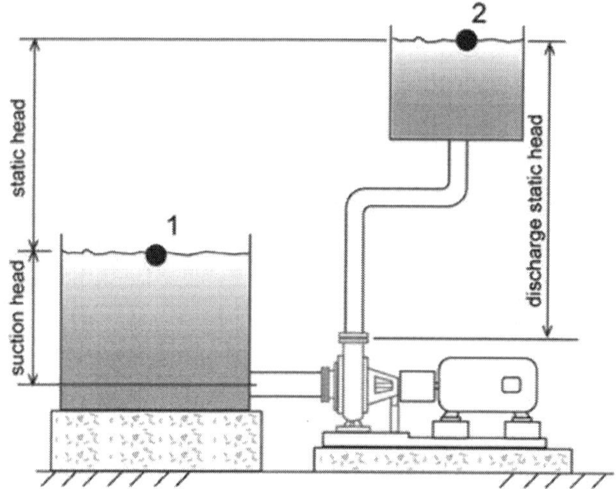

Usually the liquid is displaced from a suction tank to a discharge tank. The suction tank fluid provides pressure energy to the pump suction which helps the pump. We want to know how much pressure energy the pump itself must supply so therefore we subtract the pressure energy provided by the suction head. The static head is then the difference in height of the discharge tank fluid surface minus the suction tank fluid surface. Static head is sometimes called total static head to indicate that the pressure energy available on both sides of the pump has been considered.

Since there is a difference in height between the suction and discharge flanges or connections of a pump by convention it was agreed that the static head would be measured with respect to the suction flange elevation.

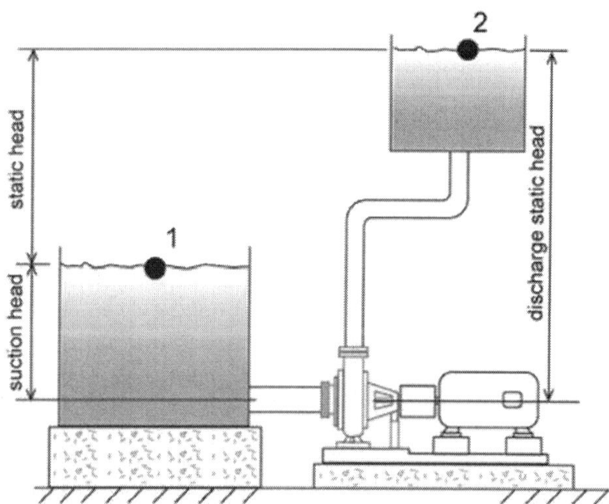

If the discharge pipe end is open to atmosphere then the static head is measured with respect to the pipe end.

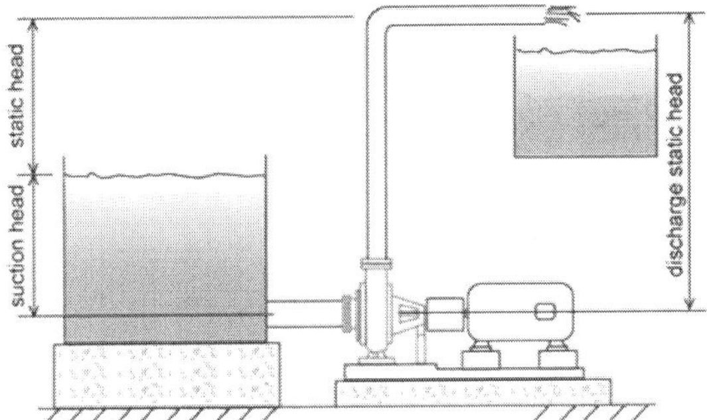

Sometimes the discharge pipe end is submerged, then the static head will be the difference in elevation between the discharge tank fluid surface and the suction tank fluid surface. Since the fluid in the system is a continuous medium and all fluid particles are connected via pressure, the fluid particles that are located at the surface of the discharge tank will contribute to the pressure built up at the pump discharge. Therefore the discharge surface elevation is the height that must be considered for static head. Avoid the mistake of using the discharge pipe end as the elevation for calculating static head if the pipe end is submerged.

Note: if the discharge pipe end is submerged, then a check valve on the pump discharge is required to avoid backflow when the pump is stopped.

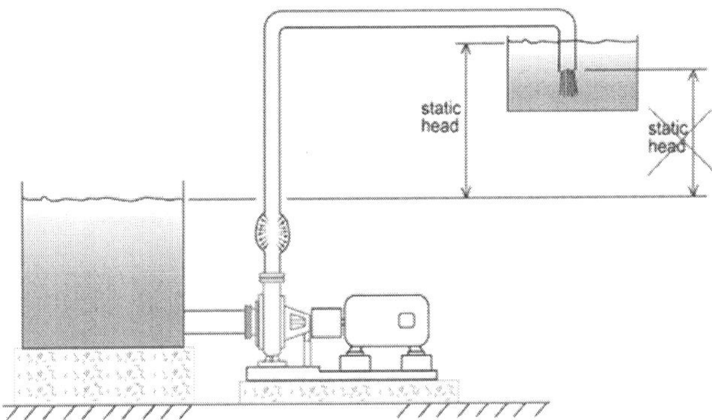

The static head can be changed by raising the surface of the discharge tank (assuming the pipe end is submerged) or suction tank or both. All of these changes will influence the flow rate.

To correctly determine the static head follow the liquid particles from start to finish, the start is almost always at the liquid surface of the suction tank, this is called the inlet elevation. The end will occur where you encounter an environment with a fixed pressure such as the open atmosphere, this point is the discharge elevation end or outlet elevation. The difference between the two elevations is the static head. The static head can be negative because the outlet elevation can be lower than the inlet elevation.

FLOW RATE DEPENDS ON ELEVATION DIFFERENCE OR STATIC HEAD

For identical systems, the flow rate will vary with the static head. If the pipe end elevation is high, the flow rate will be low. Compare this to a cyclist on a hill with a slight upward slope, his velocity will be moderate and correspond to the amount of energy he can supply to overcome the friction of the wheels on the road and the change in elevation.

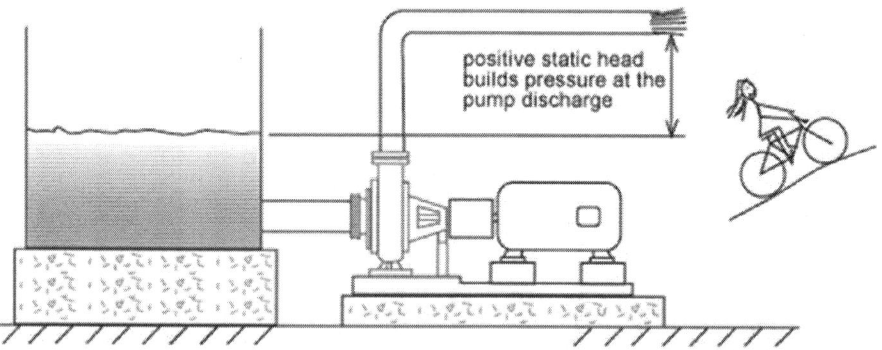

If the liquid surface of the suction tank is at the same elevation as the discharge end of the pipe then the static head will be zero and the flow rate will be limited by the friction in the system. This is equivalent to a cyclist on a flat road, his velocity depends on the amount of friction between the wheels and the road and the air resistance.

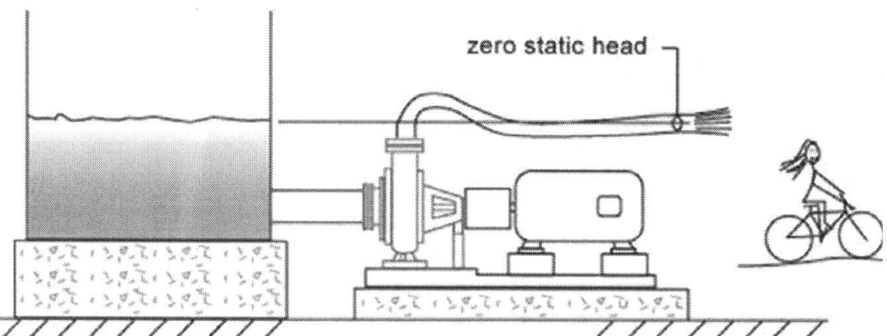

The discharge pipe end is raised vertically until the flow stops, the pump cannot raise the fluid higher than this point and the discharge pressure is at its maximum. Similarly the cyclist applies maximum force to the pedals without getting anywhere.

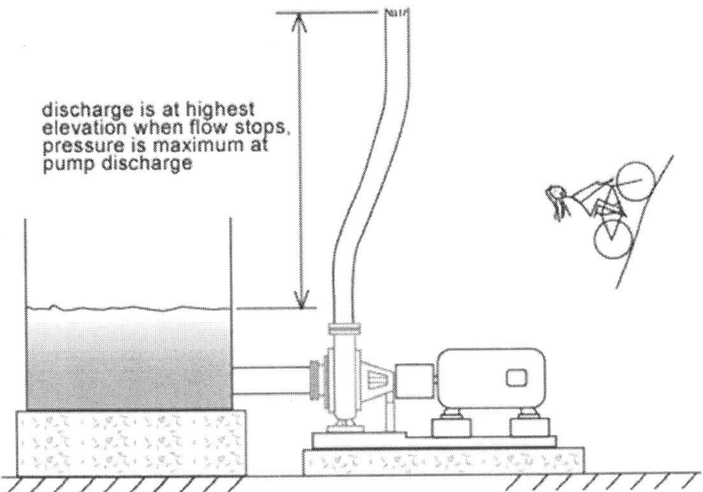

If the discharge pipe end is lower than the liquid surface of the suction tank then the static head will be negative and the flow rate high. If the negative static head is large then it is possible that a pump is not required since the energy provided by this difference in elevation may be sufficient to move the fluid through the system without the use of a pump as in the case of a siphon. By analogy, as the cyclist comes down the hill he looses his stored elevation energy which is transformed progressively into velocity energy. The lower he is on the slope, the faster he goes.

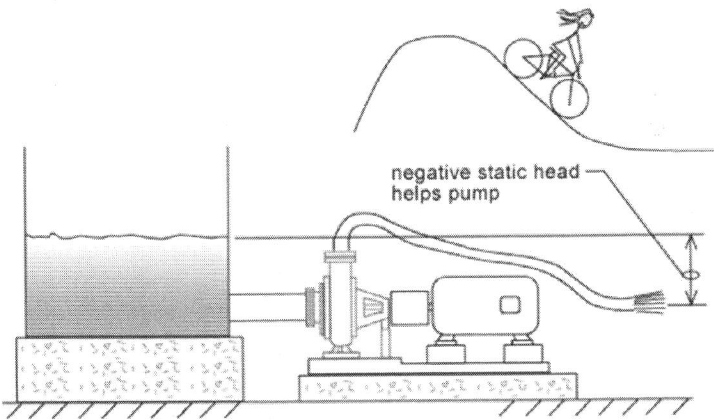

Pumps are most often rated in terms of head and flow. The discharge pipe end is raised to a height at which the flow stops, this is the head of the pump at zero flow. We measure this difference in height in feet. Head varies depending on flow rate, but in this case since there is no flow and hence no friction, the head of the pump is the maximum height that the fluid can be lifted to with respect to the surface of the suction tank. Since there is no flow the head (also called total head) that the pump produces is equal to the static head.

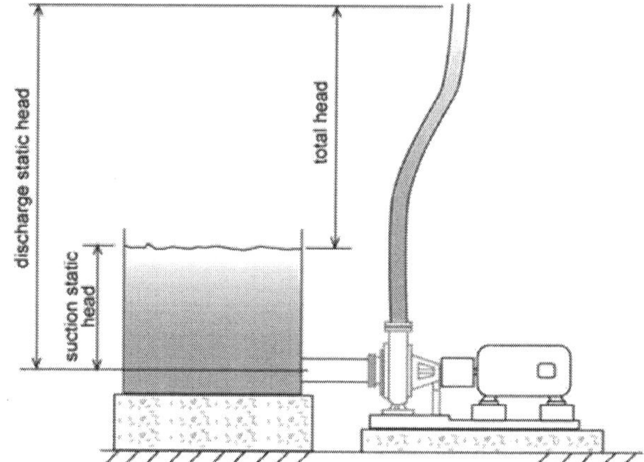

In this situation the pump will deliver its maximum pressure. The pump flow will increase and the head (also known as total head) will decrease to a value that corresponds to the flow. Why? Let's start from the point of zero flow with the pipe end at its maximum elevation, the pipe end is lowered so that flow begins. If there is flow there must be friction, the friction energy is subtracted (because it is lost) from the maximum total head and the total head is reduced. At the same time the static head is reduced which further reduces the total head.

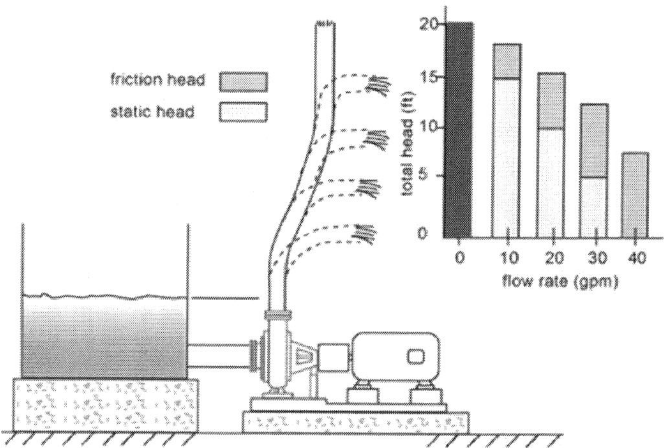

When you buy a pump you don't specify the maximum total head that the pump can deliver since this occurs at zero flow. You instead specify the total head that occurs at your required flow rate. This head will depend on the maximum height you need to reach with respect to the suction tank fluid surface and the friction loss in your system.

For example, if your pump is supplying a bathtub on the 2nd floor, you will need enough head to reach that level, that will be your static head, plus an additional amount to overcome the friction loss through the pipes and fittings.

Assuming that you want to fill the bath as quickly as possible, then the taps on the bath will be fully open and will offer very little resistance or friction loss. If you want to supply a shower head for this bathtub then you will need a pump with more head for the same flow rate because the shower head is higher and offers more resistance than the bathtub taps.

Luckily, there are many sizes and models of centrifugal pumps and you cannot expect to purchase a pump that matches exactly the head you require at the desired flow. You will probably have to purchase a pump that provides slightly more head and flow than you require and you will adjust the flow with the use of appropriate valves.

Note: you can get more head from a pump by increasing it's speed or it's impeller diameter or both. In practice, home owners cannot make these changes and to obtain a higher total head, a new pump must be purchased.

FLOW RATE DEPENDS ON FRICTION

For identical systems, the flow rate will vary with the size and diameter of the discharge pipe. A system with a discharge pipe that is generously sized will have a high flow rate. This is what happens when you put a large pipe on a tank to be emptied, it drains very fast.

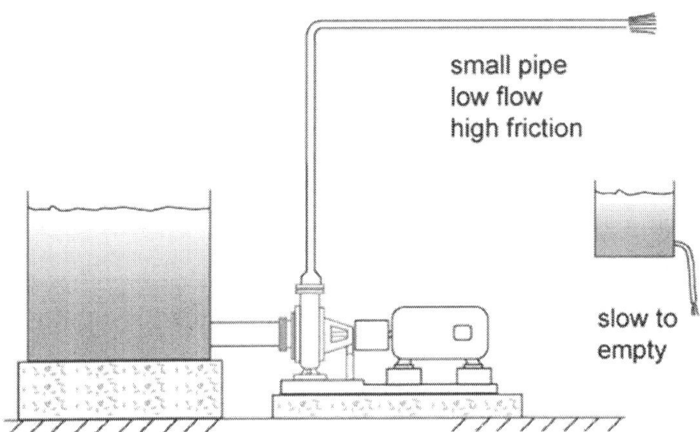

The smaller the pipe, the less the flow. How does the pump adjust itself to the diameter of the pipe, after all it does not know what size pipe will be installed? The pump you install is designed to produce a certain average flow for systems that have their pipes sized accordingly. The impeller size and its speed predispose the pump to supply the liquid at a certain flow rate. If you attempt to push that same flow through a small pipe the discharge pressure will increase and the flow will decrease. Similarly if you try to empty a tank with a small tube, it will take a long time to drain.

Later on in the tutorial, a chart will be presented giving the size of pipes for various flow rates. Or you can jump to it right away and come back later.

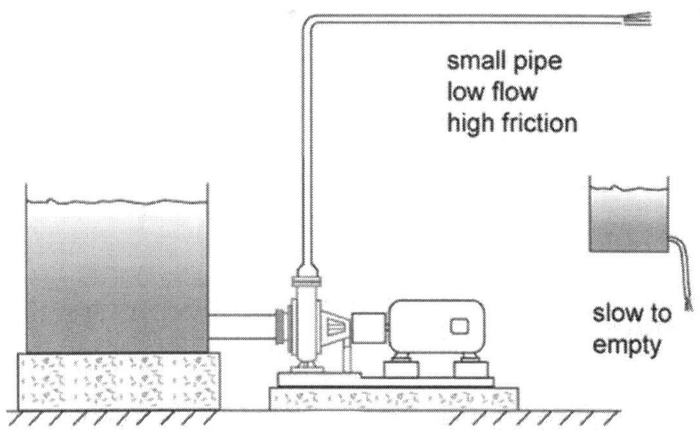

If the pipe is short the friction will be low and the flow rate high.

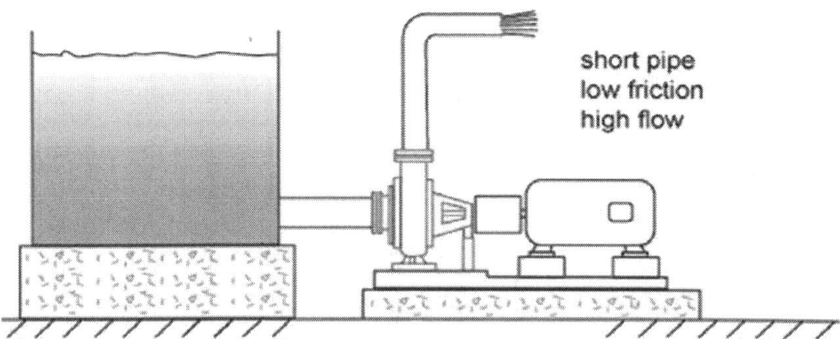

and when the discharge pipe is long, the friction will be high and the flow rate low.

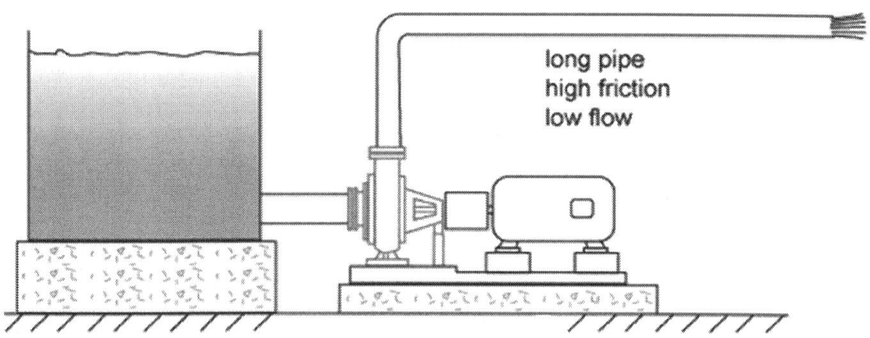

HOW DOES A CENTRIFUGAL PUMP PRODUCE PRESSURE

Fluid particles enter the pump at the suction flange or connection. They then turn 90 degrees into the impeller and fill up the volume between each impeller vane. This animation shows what happens to the fluid particles from that point forward.

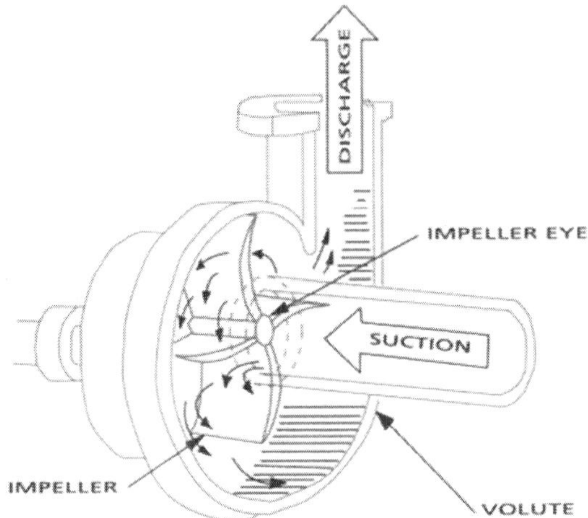

A more detail look at a more realistic cross-section of a closed impeller pump

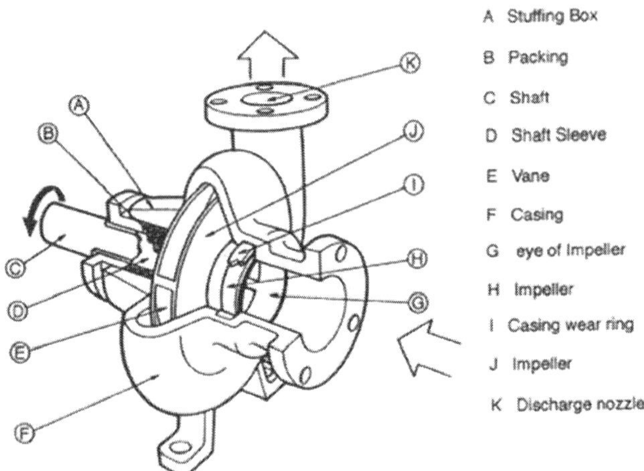

A centrifugal pump is a device whose primary purpose is to produce pressure by accelerating fluid particles to a high velocity providing them with velocity energy. What is velocity energy? It's a way to express how the velocity of objects can affect other objects, you for example. Have you ever been tackled in a football match? The velocity at which the other player comes at you determines how hard you are hit. The mass of the player is also an important factor. The combination of mass and velocity produces velocity (kinetic) energy. Another example is catching a hard baseball pitch, ouch, there can be allot of velocity in a small fast moving baseball. Fluid particles that move at high speed have velocity energy, just put your hand on the open end of a garden hose.

The fluid particles in the pump are expelled from the tips of the impeller vanes at high velocity, then they slow down as they get closer to the discharge connection, loosing some of their velocity energy. This decrease in velocity energy increases pressure energy. Unlike friction which wastes energy, the decrease in velocity energy serves to increase pressure energy, this is the principal of energy conservation in action. The same thing happens to a cyclist that starts at the top of a hill, his speed gradually increases as he looses elevation. The cyclist's elevation energy was transformed into velocity energy, in the pump's case the velocity energy is transformed into pressure energy.

Try this experiment, find a plastic cup or other container that you can poke a small pinhole in the bottom. Fill it with water and attach a string to it, and now you guessed it, start spinning it.

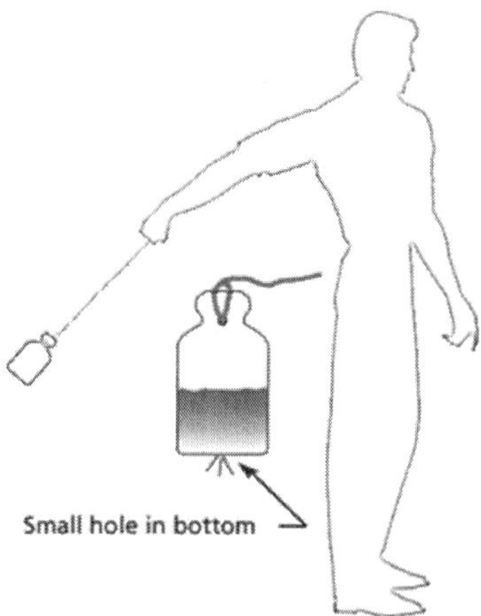

The faster you spin, the more water comes out the small hole, the water is pressurized inside the cup using centrifugal force in a similar fashion to a centrifugal pump. In the case of a pump, the rotational motion of the impeller projects fluid particles at high speed into the volume between the casing wall and the impeller tips. Prior to leaving the pump, the fluid particles slow down to the velocity at the inlet of the discharge pipe which will be the same velocity throughout the system if the pipe diameter does not change.

How does the flow rate change when the discharge pipe end elevation is changed or when there is an increase or decrease in pipe friction? These changes cause the pressure at the pump outlet to increase when the flow decreases, sounds backwards doesn't it. Well it's not and you will see why. How does the pump adjust to this change in pressure? Or in other words, if the pressure changes due to outside factors, how does the pump respond to this change.

Pressure is produced by the rotational speed of the impeller vanes. The speed is constant. The pump will produce a certain discharge pressure corresponding to the particular conditions of the system (for example, fluid viscosity, pipe size, elevation difference, etc.). If changing something in the system causes the flow to decrease (for example closing a discharge valve), there will be an increase in pressure at the pump discharge because there is no corresponding reduction in the impeller speed. The pump produces excess velocity energy because it operates at constant speed, the excess velocity energy is transformed into pressure energy and the pressure goes up.

All centrifugal pumps have a performance or characteristic curve that looks similar (assuming that the level in the suction tank remains constant), this shows how the discharge pressure varies with the flow rate through the pump.

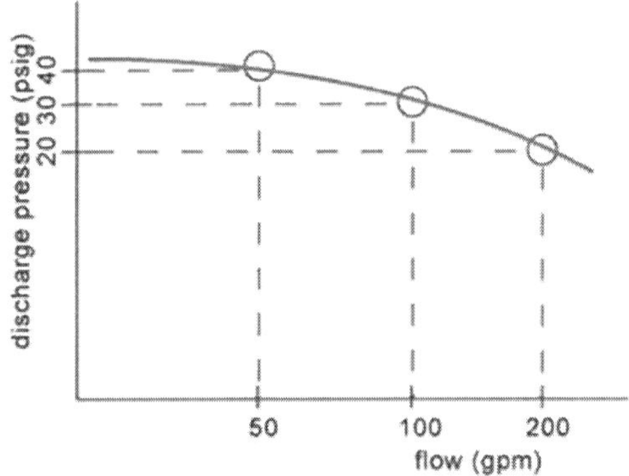

So that at 200 gpm, this pump produces 20 psig discharge pressure, and as the flow drops the pressure will reach a maximum of 40 psig.

Note: his applies to centrifugal pumps, many home owners have positive displacement pumps, often piston pumps. Those pumps produce constant flow no matter what changes are made to the system.

GENERAL OVERVIEW OF TYPES OF PUMPS ON SHIP

A ship consists of various types of fluids moving inside different machinery and systems for the purpose of cooling, heating, lubrication, and as fuels. These liquids are circulated by different types of pumps, which can be independently driven by ship power supply or attached to the machinery itself. All the systems on board ship require proper operational and compatible pump and pumping system so that ship can run on its voyage smoothly.

The selection of a type of pump for a system depends on the characteristics of the fluid to be pumped or circulated. Characteristics such as viscosity, density, surface tension and compressibility, along with characteristics of the system such as require rate of fluid, head to which the fluid is to be pumped, temperature encountered in the system, and pressure tackled by the fluid in the system, are taken into account.

TYPES OF PUMPS

The pumps used on board are broadly classified into two types:

POSITIVE DISPLACEMENT PUMP

Positive displacement pumps are self priming pumps and are normally used as priming devices.

- They consist of one or more chamber, depending upon the construction, and the chambers are alternatively filled and emptied.
- The positive displacement pumps are normally used where the discharge rate is small to medium.
- They are popularly used where the viscosity of the fluid is high.
- They are generally used to produce high pressure in the pumping system.

DYNAMIC PRESSURE OR ROTO-DYNAMIC PUMP.

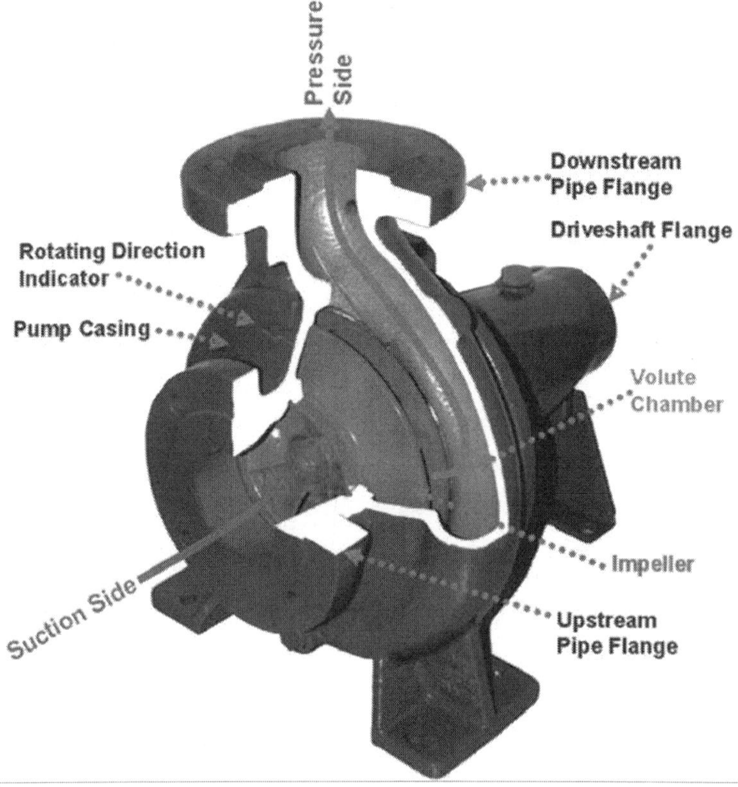

- In dynamic pressure pump, during pumping action, tangential force is imparted which accelerates the fluid normally by rotation of impeller.
- Some systems which contain dynamic pump may require positive displacement pump for priming.
- They are normally used for moderate to high discharge rate.

- The pressure differential range for this type of pumps is in a range of low to moderate.
- They are popularly used in a system where low viscosity fluids are used.

These broad classification of pumps are further differentiates by their constructional properties and popularity of usage onboard ship;

Positive Displacement pump:

- Reciprocating Pump
- Screw pump
- Gear pump
- Piston pump
- Ram type pump
- Vane pump

Dynamic pressure pumps:

- Centrifugal pumps
- Axial flow pumps
- Submersible pump
- Centrifugal-axial (mixed) pump.

THE BILGE PUMPS USED ON BOARD SHIPS

It is important to empty ship's bilge wells at regular interval of times to prevent flooding of the engine room. Bilge pumps are used for this purpose.It is important to know the construction and working of these pumps as they are used in emergency situations and also often face technical problems.

We learnt what is bilge water, oily water and use of oily water seperator. Now we will talk a look at these pumps used on board various types of ships and boats and see how does a bilge pump work. A bilge pump found on a ship is generally a reciprocating type or a centrifugal type. The rate of flow provided by a reciprocating type is lesser than that of the centrifugal type. Also, the mechanism and construction of a reciprocating pump is extremely complicated for a pump used for such important purpose.

All the bilge pumps that are generally used on ships are placed below the sea level. Though this helps them to gain extra suction pressure, they often require priming before the start. Reciprocating pumps are self-priming but due to the drawbacks mentioned earlier, their use is getting reduced day by day.

Nowadays centrifugal pumps are generally used as bilge pumps as they provide greater output rate and take lesser time for the pumping process. The only drawback of these pumps is that they don't have a self- priming system. This has not been an obstacle in their use and the problem is compensated by using an external priming system.

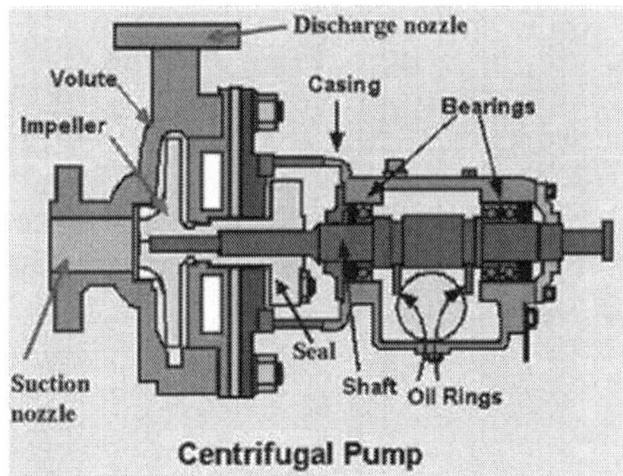

CONSTRUCTION OF BILGE PUMP

A centrifugal pump consists of an impeller. The impeller is fixed at the center with a help of a shaft.It has vanes which are fixed radially and are strategically located.

Around the impeller, a diffuser or volute is fixed. A diffuser is a ring which also has fixed blades that are strategically placed and are used to increase the speed and convert the kinetic energy of the liquid into pressure.

The impeller and diffuser are contained in a casing or the main frame which also supports the motor of the pump.

Thus there are three main parts of a centrifugal pump:

- Impeller
- Diffuser
- Casing

A sealing arrangement is provided around the shaft to prevent leaking of fluid. The sealing arrangement might consist of a gland or a mechanical seal.

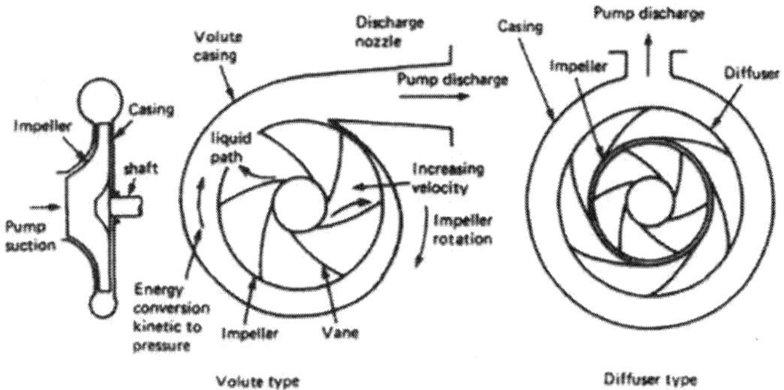

WORKING OF BILGE PUMP

The liquid enters the pump through the center of the impeller also known as the eye of the impeller. The liquid after entering flows radially out and enters the vanes of the impeller. Due to the rotation of the impeller the velocity of the liquid increases. This high velocity liquid then enters the diffuser or the volute casing where the kinetic energy of the liquid is converted to pressure. This pressurized liquid is pumped out through the pump discharge.Many centrifugal pumps have more than one impeller for increased velocity.

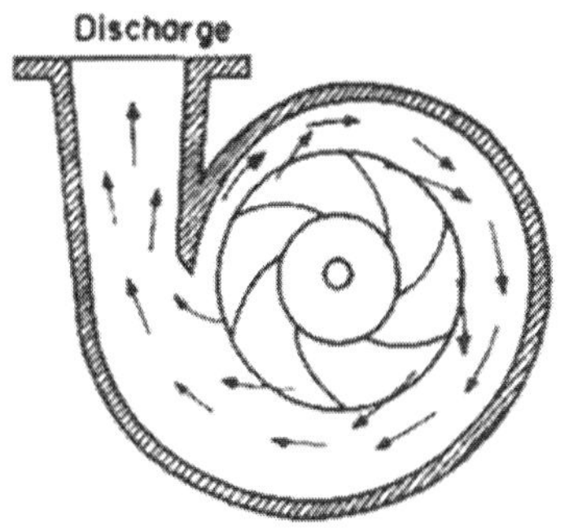

a. Volute centrifugal pump cross section

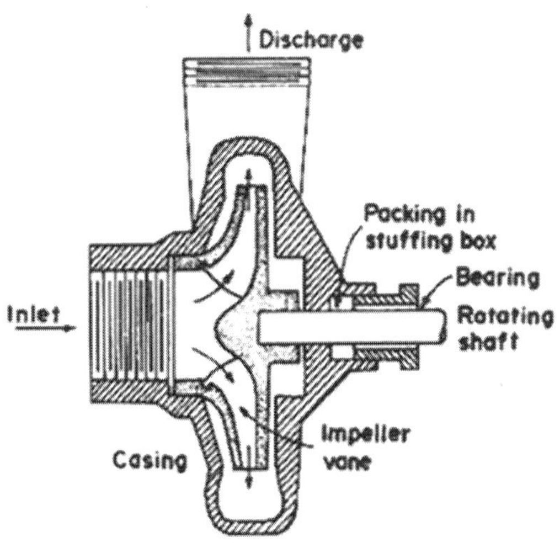

b. Horizontal centrifugal pump cross section

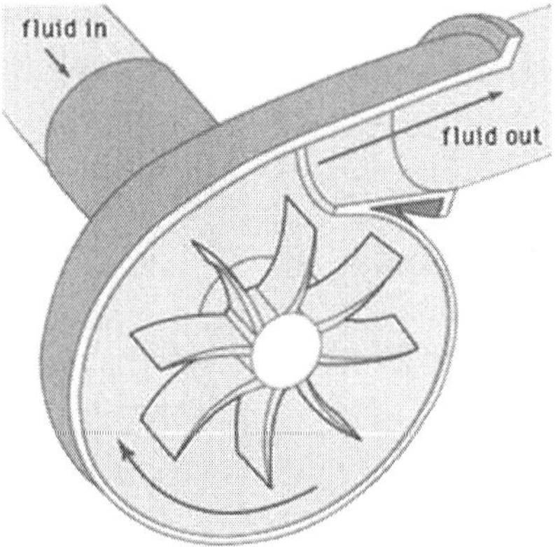

PRIMING ARRANGEMENT

As centrifugal pumps are not self priming, they require some additional arrangement to get rid of the air in the suction line. When the liquid that is to be pumped is at a level higher than the pump, then the air in the suction line can be removed by opening an air cock fitted at the pump suction. This will automatically allow the liquid to flow due to gravity.

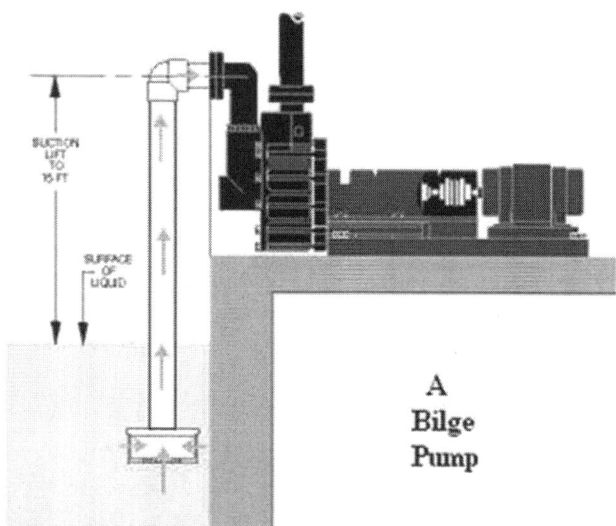

If the pump is located below the sea water level and there is a provision made to utilize the pressurized sea water, then priming can be done by opening the sea water cock and the air cock simultaneously.

Alternate priming systems can be an external air pumping unit that provides pressurised air to different pumps simultaneously.

HOW TO PRIME THE BILGE PUMP

So you want to prime a bilge pump before starting. How would you do it? Given below is the step by step procedure of doing it in the right way.

- Before starting the pump, open the suction valve and close the discharge valve.
- Start the priming unit to the suction line
- Start the motor
- Keep a watch at the priming process (It will start when you start the pump)
- Once priming is done, open the discharge valve slowly, turn by turn.
- Adjust the amount of flow with the help of the suction valve.
- Keep the desired output flow by adjusting the discharge valve
- While stopping, stop the motor first and then close the discharge and suction valve.

FRAMO HYDRAULIC CARGO PUMPING SYSTEM ON SHIPS

The framo hydraulic cargo pumping system is designed for a flexible and safe cargo and tank cleaning operation on ships. It consists of one hydraulic motor driven cargo pump installed in each cargo tank, ballast pumps, tank cleaning pumps, portable pumps and other consumers, all connected via a hydraulic ring line system to a hydraulic power unit as shown in figure below. The submerged cargo pump is a single stage centrifugal pump with the impeller close to the tank top, giving a good pumping performance of all kinds of liquids and with excellent stripping performance. The hydraulic section is surrounded by a cofferdam that completely segregates the hydraulic oil from the cargo.

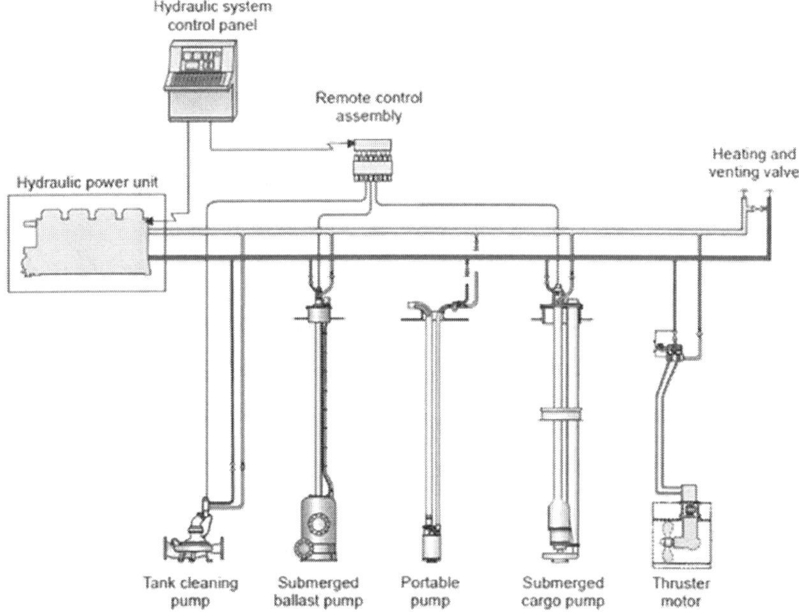

The hydraulic power unit consists of electric motor and/or diesel engine driven hydraulic power packs, where the hydraulic pumps are of axial piston type and swash plate design with variable displacement. The pump displacement is hydraulically controlled via the pressure regulator on each pump and by this system, the oil delivery from the hydraulic pumps will always be the same as the oil consumption for the hydraulic motors. To control and limit the speed of the motor, a control valve is fitted for each motor.

To keep the hydraulic oil clean and hydraulic oil temperature within desired range, a full flow filter and cooler are installed in the main return line. To regulate the oil temperature, a cooling water inlet valve is controlled from the framo control system. To prevent impurities from entering the hydraulic system, the system is also pressurized (2-6 bar) when not in operation. Depending on the installation, this is done by a jockey or feed pump.

The discharge from all cargo, ballast and other pumps connected to the system may be remotely controlled from framo control panel, vessel's integrated control system (ICS) or locally at each pump via the control valve. The portable pump is controlled locally at the pump via the control valve. The equipment is manufactured in Norway 'Framo' facility and the most important benefits of the system are:

- Segregation of cargoes for vessel's safety and the environment.
- Maximum transport volume.
- Efficient stripping and tank cleaning.
- Hydraulic system without any electrical power equipment in cargo area.
- Automatic regulation of power needed and impossible to overload.

SYSTEM DESCRIPTION

The Framo cargo pumping system is controlled by a Programmable Logic Control (PLC) installed inside the control panel. The PLC is programmed by Framo and provides the logic for safe operation of the system.

ALARM SYSTEM

All alarm inputs, except "Wear indication" are normally closed, meaning that the system is built up on normally closed contacts. Hence if a contact opens or there is a loose wire, an alarm condition occurs, i.e. FAIL TO SAFE.

The alarms are divided into two groups:

a. Shut down
b. Alarm for indication (pre-warning) only

Each alarm is indicated with a flickering light and an acoustic signal. The "Horn silence" button must be operated before it is possible to confirm the alarm by the "Acknowledge" button.

Group a) alarms: The "Reset" button must be operated to clear the alarm.

Group b) alarms: Automatic reset

POWER SUPPLY

Most hydraulic system control panels are supplied via two separate power feeders, main and back-up. In case of failure to main feeder, automatic change-over to back-up will take place. For increased availability and simplified trouble shooting, the control system is electrically divided into separate sub-systems. Each sub-system has its own 24VDC power supply. Failure in one system is not likely to interfere with other systems. The feeders and DC power supplies are monitored by relays provided with status lamps. The lamps illuminate when corresponding feeder/ supply is in good condition. A failure on any of the feeders/ supplies or short circuit will generate a potential free (dry contact) power failure alarm to the engine control room in addition to alarm at the control panel.

FEED PUMPS

The feed pumps are started/stopped manually from control panel or from the electric starter cabinet. However, when initiating start of the first main power pack, two of the feed pumps will automatically be started (high capacity mode) before the power pack is started. When the power packs are stopped, the feed pump will switch from high capacity mode to low capacity mode (only one feed pump running) automatically after 10 minutes if this is not done by operator. A running signal is provided for indication on the control panel. The high capacity mode is used to keep feed pressure on the suction side of the main hydraulic pumps. Two feed pumps must therefore be running before the main power packs can be started. If one of the feed pumps stops when running in high capacity mode and the power packs are running, the third feed pump will start automatically. If the running signal for the third feed pump is not obtained within 3 seconds, the feed pressure low alarm will be initiated and the system will be shut down. One of the feed pumps must always be running when the system is not in operation and if it stops, the protection pressure low alarm is initiated.

At system shut down, the feed pumps will stop together with the first power pack.

POWER PACKS

The power packs are started/stopped manually from control panel or from the electric starter cabinet. The power packs can be started in any sequence. Maximum 4 starts should be made during an hour. Maximum 2 following starts can be made, then 15 minutes between each. The starts are controlled by the PLC, including starts from the electric starter cabinet.

"Closed" signal from limit switches on suction line for each power pack will stop or prevent start of the corresponding power pack only. An alarm light is provided for each power pack to indicate closed valve. If more than one power pack is loaded and the hydraulic oil temperature increases to 65°C or above, the control system will automatically unload all power packs in sequence except for one. The running light will start flashing for the unloaded power packs, and the high oil temperature alarm is re-initiated each time a new power pack is unloaded. The power packs will automatically be reloaded in sequence when the hydraulic oil temperature has decreased below 60°C.

SYSTEM PRESSURE CONTROL

System pressure is set by the potentiometer on the control panel, as a voltage input to the PLC. The PLC output is amplified by a proportional valve driver card controlling the proportional pressure control valve for the system pressure. The set pressure is automatically set to zero until one of the power packs has been loaded.

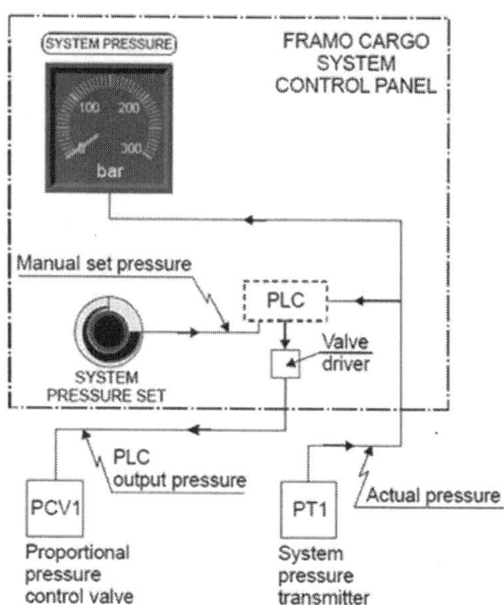

The manual set pressure is automatically compared to the actual pressure. If a pressure drop occurred, for example by starting too many consumers in proportion to number of power packs running, the PLC is limiting its output set pressure to maximum 60 bar above actual pressure. This is to avoid pressure peaks when starting another power pack or stopping a consumer. This function can be manually overruled in manual override mode 1 and is automatically overruled in mode 2.

REMOTE CONTROL OF HYDRAULICALLY DRIVEN PUMPS

The command signals from potentiometers at the front of the control panel are fed directly into the proportional valves for speed control. The built in pressure transmitters give a 4-20 mA (0-300 bar) feedback signal to the instruments on the control panel.

EMERGENCY STOP ARRANGEMENT

Following emergency stop arrangement is provided by Framo according to class requirements:

- Emergency stop push buttons located on deck are stopping the cargo pumps only, other consumers are not affected.
- Emergency stop push button located on hydraulic system control panel and central location outside engine room are stopping the hydraulic system.

SYSTEM SHUT DOWN

If a shut down function is initiated, the PLC will give shut down command to the power packs in sequence. The first power pack stops immediately and delay between shut down command of each following power pack is 0.75 seconds.

COOLING WATER INLET VALVE

The valve is automatically controlled and will open at hydraulic oil temperature above 50°C. It will stay open until the temperature decreases to below 30°C, then it will close. If no power packs are running and the hydraulic oil temperature is below 49°C, the valve will be closed. If there is a mismatch between command and feedback signal, an alarm will be released. If failure in the temperature monitoring loop, the valve will be opened.

ELECTRIC CONTROLLED BALLAST PUMP EJECTOR

The priming of ballast pump is done by creating vacuum in the pump casing. For creating vacuum, an air ejector is installed. The ejector for the priming system is automatically controlled by the PLC. The PLC receives input from a level switch installed in the pump casing and controls the solenoid valve for the ejector pilot air, starting and stopping the ejector. Time delay for start and stop is 2 seconds and the logic incorporates an interlock to prevent starting the ejector unless at least one power pack is running and operator's hydraulic pressure command is more than 30 bar.

INERT GAS PRESSURE LOW

Input from external system for shut down of cargo pumps. Visual and audible alarm provided

CARGO PRESSURE HIGH

Input from external system for shut down of cargo pumps.Visual and audible alarm provided

INTERFACE TO VESSEL'S INTEGRATED CONTROL SYSTEM (ICS)

The Framo cargo pumping system can be operated/ monitored from Framo control panel or from ICS. The mode is selected in front of the Framo control panel.

INTERFACE TO POWER MANAGEMENT SYSTEM (PMS)

A "start request" signal is sent from the Framo electric starter to PMS. Start is prohibited until a "power available" signal is received from PMS.

5

Steam Turbines and Gearing

STEAM TURBINE

Fig. The rotor of a modern steam turbine used in a power plant

A steam turbine is a device which extracts thermal energy from pressurized steam and uses it to do mechanical work on a rotating output shaft. Its modern manifestation was invented by Sir Charles Parsons in 1884.

Because the turbine generates rotary motion, it is particularly suited to be used to drive an electrical generator – about 90% of all electricity generation in the United States (1996) is by use of steam turbines. The steam turbine is a form of heat engine that derives much of its improvement in thermodynamic efficiency from the use of multiple stages in the expansion of the steam, which results in a closer approach to the ideal reversible expansion process.

HISTORY

A 250 kW industrial steam turbine from 1910 (right) directly linked to agenerator (left).

The first device that may be classified as a reaction steam turbine was little more than a toy, the classic Aeolipile, described in the 1st century by Greek mathematician Hero of Alexandria in Roman Egypt. In 1551, Taqi al-Din in Ottoman Egypt described a steam turbine with the practical application of rotating a spit. Steam turbines were also described by the Italian Giovanni Branca (1629) andJohn Wilkins in England (1648). The devices described by Taqi al-Din and Wilkins are today known as steam jacks.

The modern steam turbine was invented in 1884 by Sir Charles Parsons, whose first model was connected to a dynamo that generated 7.5 kW (10 hp) of electricity. The invention of Parsons' steam turbine made cheap and plentiful electricity possible and revolutionized marine transport and naval warfare. Parsons' design was a reaction type. His patent was licensed and the turbine scaled-up shortly after by an American, George Westinghouse. The Parsons turbine also turned out to be easy to scale up. Parsons had the satisfaction of seeing his invention adopted for all major world power stations, and the size of generators had increased from his first 7.5 kW set up to units of 50,000 kW capacity. Within Parson's lifetime, the generating capacity of a unit was scaled up by about 10,000 times, and the total output from turbo-generators constructed by his firm C. A. Parsons and Company and by their licensees, for land purposes alone, had exceeded thirty million horse-power.

A number of other variations of turbines have been developed that work effectively with steam. The *de Laval turbine* (invented by Gustaf de Laval) accelerated the steam to full speed before running it against a turbine blade. De Laval's impulse turbine is simpler, less expensive and does not need to be pressure-proof.

It can operate with any pressure of steam, but is considerably less efficient. fr:Auguste Rateau developed a pressure compounded impulse turbine using the de Laval principle as early as 1896,obtained a US patent in 1903, and applied the turbine to a French torpedo boat in 1904. He taught at the École des mines de Saint-Étienne for a decade until 1897, and later founded a successful company that was incorporated into the Alstom firm after his death. One of the founders of the modern theory of steam and gas turbines was Aurel Stodola, a Slovak physicist and engineer and professor at the Swiss Polytechnical Institute (now ETH) in Zurich. His work *Die Dampfturbinen und ihre Aussichten als Wärmekraftmaschinen* (English: The Steam Turbine and its prospective use as a Mechanical Engine) was published in Berlin in 1903. A further book *Dampf und Gas-Turbinen*(English: Steam and Gas Turbines) was published in 1922.

The *Brown-Curtis turbine*, an impulse type, which had been originally developed and patented by the U.S. company International Curtis Marine Turbine Company, was developed in the 1900s in conjunction with John Brown & Company. It was used in John Brown-engined merchant ships and warships, including liners and Royal Navy warships.

MANUFACTURING

The manufacturing industry for steam turbines is dominated by Chinese power equipment makers. Harbin Electric, Shanghai Electric, and Dongfang Electric, the top three power equipment makers in China collectively hold a majority stake in the worldwide market share for steam turbines in 2009-10 according to Platts. Other manufacturers with minor market share include Bhel, Siemens, Alstom, GE, Mitsubishi Heavy Industries, and Toshiba. The consulting firm Frost & Sullivan projects that manufacturing of steam turbines will become more consolidated by 2020 as Chinese power manufacturers win increasing business outside of China.

TYPES

Steam turbines are made in a variety of sizes ranging from small <0.75 kW (<1 hp) units (rare) used as mechanical drives for pumps, compressors and other shaft driven equipment, to 1 500 000 kW (1.5 GW; 2 000 000 hp) turbines used to generate electricity. There are several classifications for modern steam turbines.

Blade and stage design

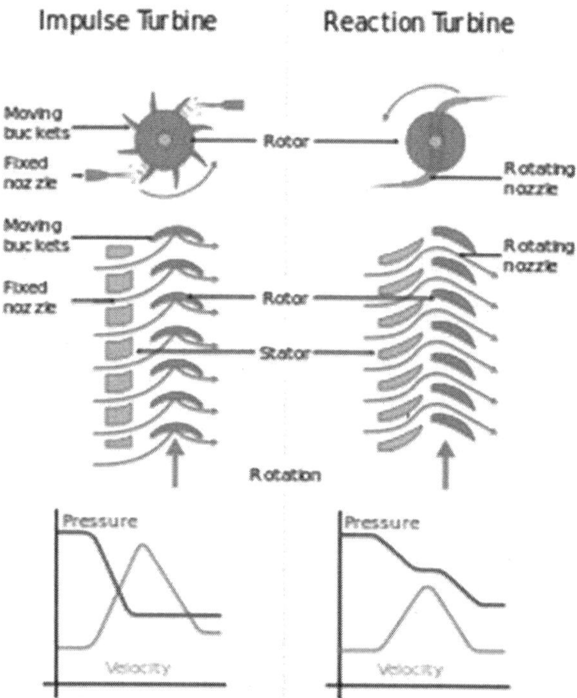

Schematic diagram outlining the difference between an impulse and a 50% reaction turbine

Turbine blades are of two basic types, blades and nozzles. Blades move entirely due to the impact of steam on them and their profiles do not converge. This results in a steam velocity drop and essentially no pressure drop as steam moves through the blades. A turbine composed of blades alternating with fixed nozzles is called an impulse turbine, Curtis turbine, Rateau turbine, or Brown-Curtis turbine. Nozzles appear similar to blades, but their profiles converge near the exit. This results in a steam pressure drop and velocity increase as steam moves through the nozzles. Nozzles move due to both the impact of steam on them and the reaction due to the high-velocity steam at the exit. A turbine composed of moving nozzles alternating with fixed nozzles is called a reaction turbineor Parsons turbine.

Except for low-power applications, turbine blades are arranged in multiple stages in series, called compounding, which greatly improves efficiency at low speeds. A reaction stage is a row of fixed nozzles followed by a row of moving nozzles. Multiple reaction stages divide the pressure drop between the steam inlet and exhaust into numerous small drops, resulting in a pressure-compounded turbine. Impulse stages may be either pressure-compounded, velocity-compounded, or pressure-velocity compounded. A pressure-compounded impulse stage is a row of fixed nozzles followed by a row of moving blades, with multiple stages for compounding. This is also known as a Rateau turbine, after its inventor. A velocity-compounded impulse stage (invented by Curtis and also called a "Curtis wheel") is a row of fixed nozzles followed by two or more rows of moving blades alternating with rows of fixed blades. This divides the velocity drop across the stage into several smaller drops. A series of velocity-compounded impulse stages is called a pressure-velocity compounded turbine.

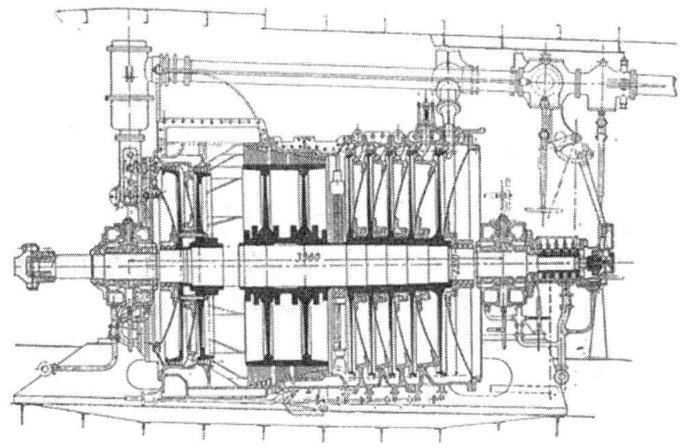

Diagram of an AEG marine steam turbine circa 1905

By 1905, when steam turbines were coming into use on fast ships (such asHMS *Dreadnought*) and in land-based power applications, it had been determined that it was desirable to use one or more Curtis wheels at the beginning of a multi-stage turbine (where the steam pressure is highest), followed by reaction stages. This was more efficient with high-pressure steam due to reduced leakage between the turbine rotor and the casing. This is illustrated in the drawing of the German 1905 AEG marine steam turbine. The steam from the boilers enters from the right at high pressure through a throttle, controlled manually by an operator (in this case a sailor known as the throttleman).

It passes through five Curtis wheels and numerous reaction stages (the small blades at the edges of the two large rotors in the middle) before exiting at low pressure, almost certainly to a condenser. The condenser provides a vacuum that maximizes the energy extracted from the steam, and condenses the steam into feedwater to be returned to the boilers. On the left are several additional reaction stages (on two large rotors) that rotate the turbine in reverse for astern operation, with steam admitted by a separate throttle. Since ships are rarely operated in reverse, efficiency is not a priority in astern turbines, so only a few stages are used to save cost.

Blade Design Challenges

A major challenge facing turbine design is reducing the creep experienced by the blades. Because of the high temperatures and high stresses of operation, steam turbine materials become damaged through these mechanisms. As temperatures are increased in an effort to improve turbine efficiency, creep becomes more significant. To limit creep, thermal coatings and superalloys with solid-solution strengthening and grain boundary strengthening are used in blade designs.

Protective coatings are used to reduce the thermal damage and to limit oxidation. These coatings are often stabilized zirconium dioxide-based ceramics. Using a thermal protective coating limits the temperature exposure of the nickel superalloy. This reduces the creep mechanisms experienced in the blade. Oxidation coatings limit efficiency losses caused by a buildup on the outside of the blades, which is especially important in the high-temperature environment.

The nickel-based blades are alloyed with aluminum and titanium to improve strength and creep resistance. The microstructure of these alloys is composed of different regions of composition. A uniform dispersion of the gamma-prime phase – a combination of nickel, aluminum, and titanium – promotes the strength and creep resistance of the blade due to the microstructure.

Refractory elements such as rhenium and ruthenium can be added to the alloy to improve creep strength. The addition of these elements reduces the diffusion of the gamma prime phase, thus preserving the fatigue resistance, strength, and creep resistance.

Steam supply and exhaust conditions

A low-pressure steam turbine in a nuclear power plant. These turbines exhaust steam at a pressure below atmospheric.

These types include condensing, non-condensing, reheat, extraction and induction.

Condensing turbines are most commonly found in electrical power plants. These turbines receive steam from a boiler and exhaust it to acondenser. The exhausted steam is at a pressure well below atmospheric, and is in a partially condensed state, typically of a quality near 90%.

Non-condensing or back pressure turbines are most widely used for process steam applications. The exhaust pressure is controlled by a regulating valve to suit the needs of the process steam pressure. These are commonly found at refineries, district heating units, pulp and paper plants, and desalination facilities where large amounts of low pressure process steam are needed.

Reheat turbines are also used almost exclusively in electrical power plants. In a reheat turbine, steam flow exits from a high pressure section of the turbine and is returned to the boiler where additional superheat is added. The steam then goes back into an intermediate pressure section of the turbine and continues its expansion. Using reheat in a cycle increases the work output from the turbine and also the expansion reaches conclusion before the steam condenses, thereby minimizing the erosion of the blades in last rows. In most of the cases, maximum number of reheats employed in a cycle is 2 as the cost of super-heating the steam negates the increase in the work output from turbine.

Extracting type turbines are common in all applications. In an extracting type turbine, steam is released from various stages of the turbine, and used for industrial process needs or sent to boiler feedwater heaters to improve overall cycle efficiency. Extraction flows may be controlled with a valve, or left uncontrolled.

Induction turbines introduce low pressure steam at an intermediate stage to produce additional power.

Casing or shaft arrangements

These arrangements include single casing, tandem compound and cross compound turbines. Single casing units are the most basic style where a single casing and shaft are coupled to a generator. Tandem compound are used where two or more casings are directly coupled together to drive a single generator. A cross compound turbine arrangement features two or more shafts not in line driving two or more generators that often operate at different speeds. A cross compound turbine is typically used for many large applications.

Two-flow rotors

A two-flow turbine rotor. The steam enters in the middle of the shaft, and exits at each end, balancing the axial force.

The moving steam imparts both a tangential and axial thrust on the turbine shaft, but the axial thrust in a simple turbine is unopposed. To maintain the correct rotor position and balancing, this force must be counteracted by an opposing force. Thrust bearings can be used for the shaft bearings, the rotor can use dummy pistons, it can be double flow- the steam enters in the middle of the shaft and exits at both ends, or a combination of any of these. In a double flow rotor, the blades in each half face opposite ways, so that the axial forces negate each other but the tangential forces act together. This design of rotor is also called two-flow, double-axial-flow, or double-exhaust. This arrangement is common in low-pressure casings of a compound turbine.

PRINCIPLE OF OPERATION AND DESIGN

A simple turbine schematic of the Parsons type: rotating and fixed stators alternate and steam pressure drops by a fraction of the total across each pair. The stators grow larger as pressure drops.

An ideal steam turbine is considered to be an isentropic process, or constant entropy process, in which the entropy of the steam entering the turbine is equal to the entropy of the steam leaving the turbine. No steam turbine is truly isentropic, however, with typical isentropic efficiencies ranging from 20–90% based on the application of the turbine. The interior of a turbine comprises several sets of blades or*buckets*. One set of stationary blades is connected to the casing and one set of rotating blades is connected to the shaft. The sets intermesh with certain minimum clearances, with the size and configuration of sets varying to efficiently exploit the expansion of steam at each stage.

TURBINE EFFICIENCY

To maximize turbine efficiency the steam is expanded, doing work, in a number of stages. These stages are characterized by how the energy is extracted from them and are known as either impulse or reaction turbines. Most steam turbines use a mixture of the reaction and impulse designs: each stage behaves as either one or the other, but the overall turbine uses both. Typically, higher pressure sections are reaction type and lower pressure stages are impulse type.

Impulse turbines

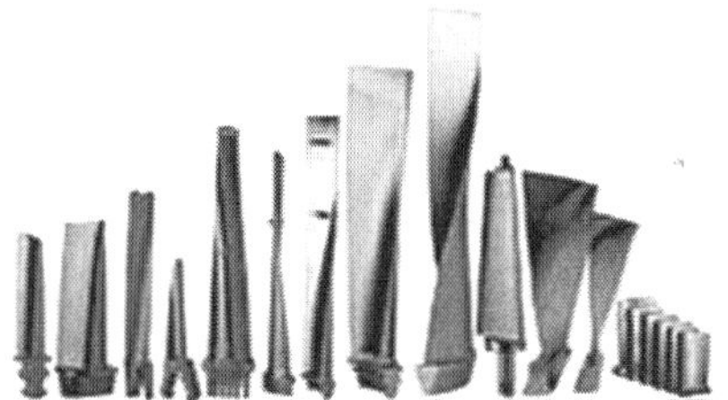

A selection of impulse turbine blades

An impulse turbine has fixed nozzles that orient the steam flow into high speed jets. These jets contain significant kinetic energy, which is converted into shaft rotation by the bucket-like shaped rotor blades, as the steam jet changes direction. A pressure drop occurs across only the stationary blades, with a net increase in steam velocity across the stage. As the steam flows through the nozzle its pressure falls from inlet pressure to the exit pressure (atmospheric pressure, or more usually, the condenser vacuum). Due to this high ratio of expansion of steam, the steam leaves the nozzle with a very high velocity. The steam leaving the moving blades has a large portion of the maximum velocity of the steam when leaving the nozzle. The loss of energy due to this higher exit velocity is commonly called the carry over velocity or leaving loss.

The law of moment of momentum states that the sum of the moments of external forces acting on a fluid which is temporarily occupying the control volume is equal to the net time change of angular momentum flux through the control volume.

The swirling fluid enters the control volume at radius r_1 with tangential velocity V_{w1} and leaves at radius r_2 with tangential velocity V_{w2}.

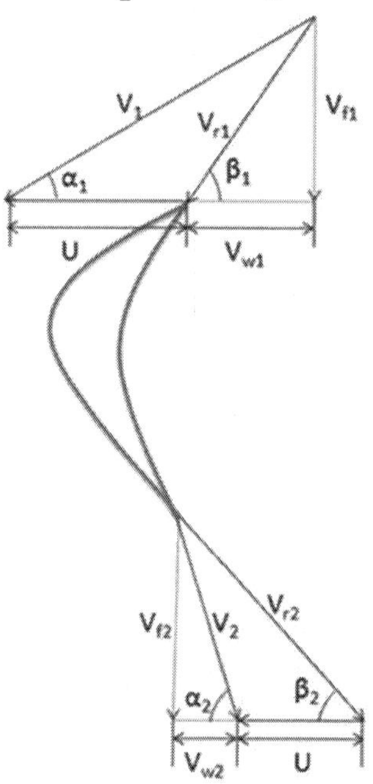

Velocity triangle

A velocity triangle paves the way for a better understanding of the relationship between the various velocities. In the adjacent figure we have:

V_1 and V_2 are the absolute velocities at the inlet and outlet respectively.
V_{f1} and V_{f2} are the flow velocities at the inlet and outlet respectively.
V_{w1} + U and V_{w2} are the swirl velocities at the inlet and outlet respectively.
V_{r1} and V_{r2} are the relative velocities at the inlet and outlet respectively.
U_1 and U_2 are the velocities of the blade at the inlet and outlet respectively.
α is the guide vane angle and β is the blade angle.

Then by the law of moment of momentum, the torque on the fluid is given by:

$$T = \dot{m}(r_2 V_{w2} - r_1 V_{w1})$$

For an impulse steam turbine: $r_2 = r_1 = r$. Therefore, the tangential force on the blades is$F_u = \dot{m}(V_{w1} - V_{w2})$. The work done per unit time or power developed: $W = T^*\omega$.

When ω is the angular velocity of the turbine, then the blade speed is $U = \omega * r$. The power developed is then $W = \dot{m}U(\Delta V_w)$.

Blade efficiency

Blade efficiency (η_b) can be defined as the ratio of the work done on the blades to kinetic energy supplied to the fluid, and is given by

$$\eta_b = \frac{Work\ Done}{Kinetic\ Energy\ Supplied} = \frac{2UV_w}{V_1^2}$$

Stage efficiency

Convergent-divergent nozzle

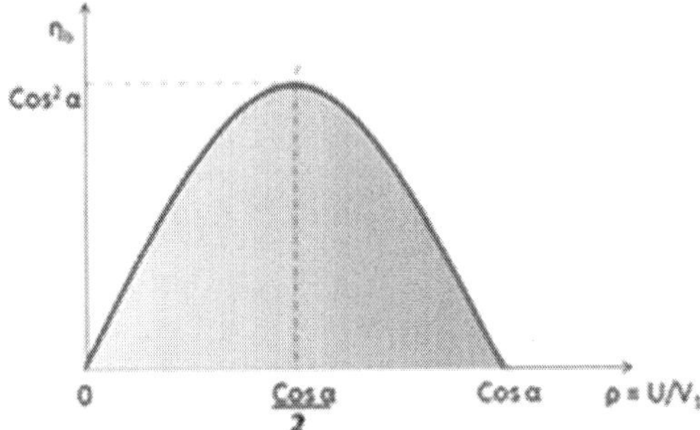

Graph depicting efficiency of Impulse turbine

A stage of an impulse turbine consists of a nozzle set and a moving wheel. The stage efficiency defines a relationship between enthalpy drop in the nozzle and work done in the stage.

$$\eta_{stage} = \frac{Work\ done\ on\ blade}{Energy\ supplied\ per\ stage} = \frac{U\Delta V_w}{\Delta h}$$

Where $\Delta h = h_2 - h_1$ is the specific enthalpy drop of steam in the nozzle. By the first law of thermodynamics:

$$h_1 + \frac{V_1^2}{2} = h_2 + \frac{V_2^2}{2}$$

Assuming that V_1 is appreciably less than V_2, we get

$$\Delta h \approx \frac{V_2^2}{2}$$

Furthermore, stage efficiency is the product of blade efficiency and nozzle efficiency, or $\eta_{stage} = \eta_b * \eta_N$.

Nozzle efficiency is given by,

$$\eta_N = \frac{V_2^2}{2(h_1 - h_2)},$$

where the enthalpy (in J/Kg) of steam at the entrance of the nozzle is h_1 and the enthalpy of steam at the exit of the nozzle is,

$$h_2. \Delta V_w = V_{w1} - (-V_{w2}) \, \Delta V_w = V_{w1} + V_{w2}.$$

$$\Delta V_w = V_{r1}\cos\beta_1 + V_{r2}\cos\beta_2 \, \Delta V_w = V_{r1}\cos\beta_1 (1 + \frac{V_{r2}\cos\beta_2}{V_{r1}\cos\beta_1})$$

The ratio of the cosines of the blade angles at the outlet and inlet can be taken and denoted,

$$c = \frac{\cos\beta_2}{\cos\beta_1}.$$

The ratio of steam velocities relative to the rotor speed at the outlet to the inlet of the blade is defined by the friction coefficient

$$k = \frac{V_{r2}}{V_{r1}}.$$

$k < 1$ and depicts the loss in the relative velocity due to friction as the steam flows around the blades ($k = 1$ for smooth blades).

$$\eta_b = \frac{2U\Delta V_w}{V_1^2} = \frac{2U(\cos\alpha_1 - U/V_1)(1 + kc)}{V_1}$$

The ratio of the blade speed to the absolute steam velocity at the inlet is termed the blade speed ratio,

$$\rho = \frac{U}{V_1}$$

η_b is maximum when,

$$\frac{d\eta_b}{d\rho} = 0$$

or,

$$\frac{d}{d\rho}(2\cos\alpha_1 - \rho^2(1+kc)) = 0.$$

That implies,

$$\rho = \frac{\cos\alpha_1}{2}$$

and therefore,

$$\frac{U}{V_1} = \frac{\cos\alpha_1}{2}.$$

Now,

$$\rho_{opt} = \frac{U}{V_1} = \frac{\cos\alpha_1}{2}$$

(for a single stage impulse turbine)

Therefore, the maximum value of stage efficiency is obtained by putting the value of,

$$\frac{U}{V_1} = \frac{\cos\alpha_1}{2}$$

in the expression of η_b/

We get:

$$(\eta_b)_{max} = 2(\rho\cos\alpha_1 - \rho^2)(1+kc) = \frac{\cos^2\alpha_1(1+kc)}{2}.$$

For equiangular blades, $\beta_1 = \beta_2$, therefore $c = 1$, and we get,

$$(\eta_b)_{max} = \frac{\cos^2\alpha_1(1+k)}{2}.$$

If the friction due to the blade surface is neglected then,

$$(\eta_b)_{max} = \cos^2\alpha_1.$$

CONCLUSIONS ON MAXIMUM EFFICIENCY

$$(\eta_b)_{max} = \cos^2\alpha_1$$

1. For a given steam velocity work done per kg of steam would be maximum when $\cos^2\alpha_1 = 1$ or $\alpha_1 = 0$.
2. As α_1 increases, the work done on the blades reduces, but at the same time surface area of the blade reduces, therefore there are less frictional losses.

Reaction turbines

In the *reaction turbine*, the rotor blades themselves are arranged to form convergent nozzles. This type of turbine makes use of the reaction force produced as the steam accelerates through the nozzles formed by the rotor. Steam is directed onto the rotor by the fixed vanes of the stator. It leaves the stator as a jet that fills the entire circumference of the rotor. The steam then changes direction and increases its speed relative to the speed of the blades. A pressure drop occurs across both the stator and the rotor, with steam accelerating through the stator and decelerating through the rotor, with no net change in steam velocity across the stage but with a decrease in both pressure and temperature, reflecting the work performed in the driving of the rotor.

BLADE EFFICIENCY

Energy input to the blades in a stage:

$E = \Delta h$ is equal to the kinetic energy supplied to the fixed blades (f) + the kinetic energy supplied to the moving blades (m).

Or, E = enthalpy drop over the fixed blades, Δh_f + enthalpy drop over the moving blades, Δh_m.

The effect of expansion of steam over the moving blades is to increase the relative velocity at the exit. Therefore, the relative velocity at the exit V_{r2} is always greater than the relative velocity at the inlet V_{r1}.

In terms of velocities, the enthalpy drop over the moving blades is given by:

$$\Delta h_m = \frac{V_{r2}^2 - V_{r1}^2}{2}$$

(it contributes to a change in static pressure)

The enthalpy drop in the fixed blades, with the assumption that the velocity of steam entering the fixed blades is equal to the velocity of steam leaving the previously moving blades is given by:

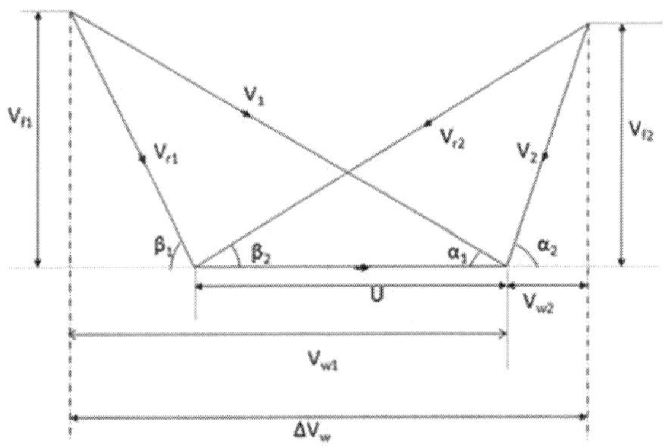

Velocity diagram,

$$\Delta h_f = \frac{V_1^2 - V_0^2}{2}$$

where V_0 is the inlet velocity of steam in the nozzle

V_0 is very small and hence can be neglected

Therefore,

$$\Delta h_f = \frac{V_1^2}{2}$$

$$E = \Delta h_f + \Delta h_m$$

$$E = \frac{V_1^2}{2} + \frac{V_{r2}^2 - V_{r1}^2}{2}$$

A very widely used design has half degree of reaction or 50% reaction and this is known as Parson's turbine. This consists of symmetrical rotor and stator blades. For this turbine the velocity triangle is similar and we have:

$$\alpha_1 = \beta_2, \beta_1 = \alpha_2$$

$$V_1 = V_{r2}, V_{r1} = V_2$$

Assuming *Parson's turbine* and obtaining all the expressions we get

$$E = V_1^2 - \frac{V_{r1}^2}{2}$$

From the inlet velocity triangle we have,

$$V_{r1}^2 = V_1^2 + U^2 - 2UV_1 \cos\alpha_1$$

$$E = V_1^2 - \frac{V_1^2}{2} - \frac{U^2}{2} + \frac{2UV_1 \cos\alpha_1}{2}$$

$$E = \frac{V_1^2 - U^2 + 2UV_1 \cos\alpha_1}{2}$$

Work done (for unit mass flow per second):

$$W = U * \Delta V_w = U * (2 * V_1 \cos\alpha_1 - U)$$

Therefore, the blade efficiency is given by

$$\eta_b = \frac{2U(2V_1 \cos\alpha_1 - U)}{V_1^2 - U^2 + 2V_1 U \cos\alpha_1}$$

CONDITION OF MAXIMUM BLADE EFFICIENCY

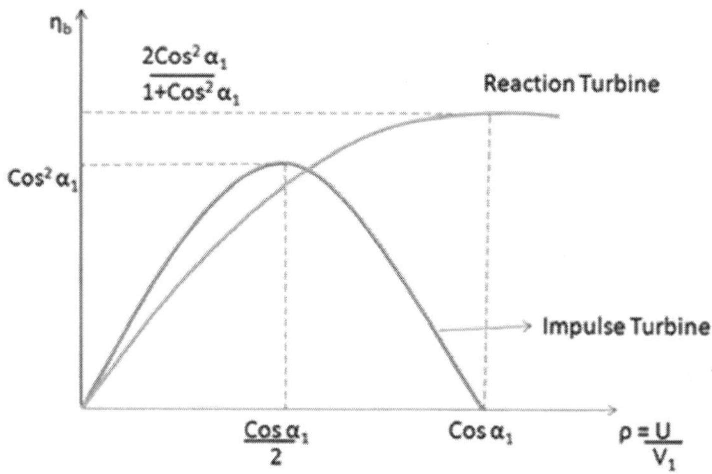

Comparing Efficiencies of Impulse and Reaction turbines

If $\rho = \frac{U}{V_1}$, then

$$(\eta_b)_{max} = \frac{2\rho(\cos\alpha_1 - \rho)}{V_1^2 - U^2 + 2UV_1\cos\alpha_1}$$

For maximum efficiency $\frac{d\eta_b}{d\rho} = 0$, we get,

$$(1 - \rho^2 + 2\rho\cos\alpha_1)(4\cos\alpha_1 - 4\rho) - 2\rho(2\cos\alpha_1 - \rho)(-2\rho + 2\cos\alpha_1) = 0$$

and this finally gives,

$$\rho_{opt} = \frac{U}{V_1} = \cos\alpha_1$$

Therefore, $(\eta_b)_{max}$ is found by putting the value of $\rho = \cos\alpha_1$ in the expression of blade efficiency,

$$(\eta_b)_{reaction} = \frac{2\cos^2\alpha_1}{1 + \cos^2\alpha_1}$$

$$(\eta_b)_{impulse} = \cos^2\alpha_1$$

Operation and maintenance

Because of the high pressures used in the steam circuits and the materials used, steam turbines and their casings have high thermal inertia. When warming up a steam turbine for use, the main steam stop valves (after the boiler) have a bypass line to allow superheated steam to slowly bypass the valve and proceed to heat up the lines in the system along with the steam turbine. Also, a turning gear is engaged when there is no steam to slowly rotate the turbine to ensure even heating to prevent uneven expansion. After first rotating the turbine by the turning gear, allowing time for the rotor to assume a straight plane (no bowing), then the turning gear is disengaged and steam is admitted to the turbine, first to the astern blades then to the ahead blades slowly rotating the turbine at 10–15 RPM (0.17–0.25 Hz) to slowly warm the turbine. The warm up procedure for large steam turbines may exceed ten hours.

A modern steam turbine generator installation

During normal operation, rotor imbalance can lead to vibration, which, because of the high rotation velocities, could lead to a blade breaking away from the rotor and through the casing. To reduce this risk, considerable efforts are spent to balance the turbine. Also, turbines are run with high quality steam: either superheated (dry) steam, or saturated steam with a high dryness fraction. This prevents the rapid impingement and erosion of the blades which occurs when condensed water is blasted onto the blades (moisture carry over). Also, liquid water entering the blades may damage the thrust bearings for the turbine shaft. To prevent this, along with controls and baffles in the boilers to ensure high quality steam, condensate drains are installed in the steam piping leading to the turbine.

Maintenance requirements of modern steam turbines are simple and incur low costs (typically around $0.005 per kWh); their operational life often exceeds 50 years.

Speed regulation

The control of a turbine with a governor is essential, as turbines need to be run up slowly to prevent damage and some applications (such as the generation of alternating current electricity) require precise speed control. Uncontrolled acceleration of the turbine rotor can lead to an overspeed trip, which causes the nozzle valves that control the flow of steam to the turbine to close. If this fails then the turbine may continue accelerating until it breaks apart, often catastrophically. Turbines are expensive to make, requiring precision manufacture and special quality materials.

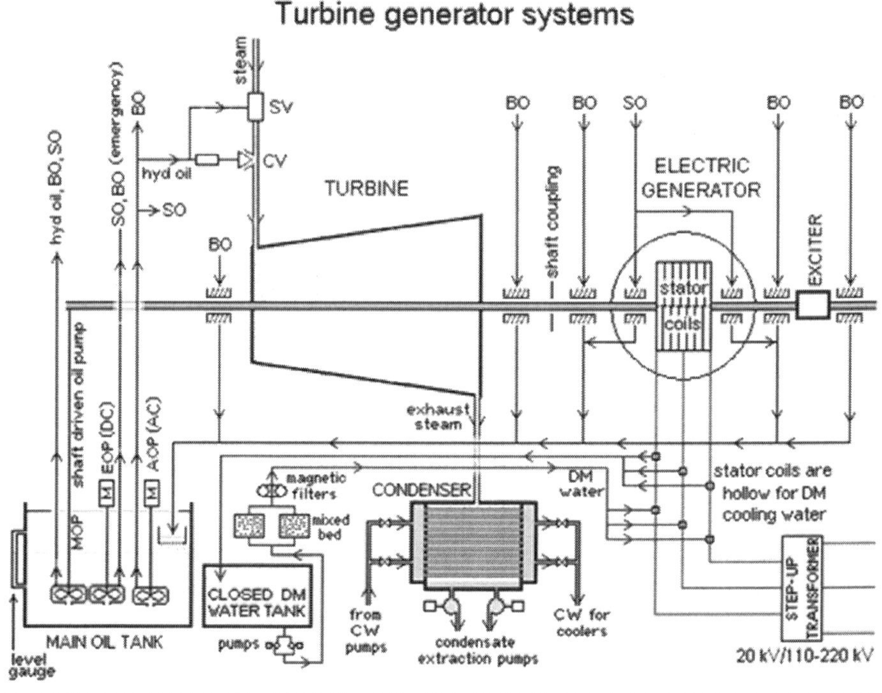

SO-seal oil; BO-bearing oil; hyd oil-hydraulic oil; SV-stop valve; CV-control valve;
MOP-main oil pump; EOP-emergency oil pump; AOP-auxiliary oil pump; [M] - motor
CW-circulating water; DM-demineralised (water); DC - direct current; AC - alternating current

Diagram of a steam turbine generator system

During normal operation in synchronization with the electricity network, power plants are governed with a five percent droop speed control. This means the full load speed is 100% and the no-load speed is 105%. This is required for the stable operation of the network without hunting and drop-outs of power plants. Normally the changes in speed are minor. Adjustments in power output are made by slowly raising the droop curve by increasing the spring pressure on a centrifugal governor. Generally this is a basic system requirement for all power plants because the older and newer plants have to be compatible in response to the instantaneous changes in frequency without depending on outside communication.

Thermodynamics of steam turbines

The steam turbine operates on basic principles of thermodynamics using the part 3-4 of the Rankine cycle shown in the adjoining diagram. Superheated steam (or dry saturated steam, depending on application) leaves the boiler at high temperature and high pressure.

T-s diagram for steam

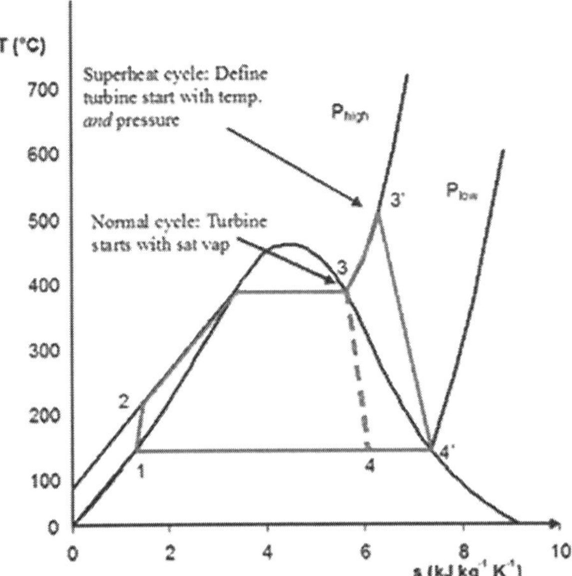

T-s diagram of a superheated Rankine cycle

At entry to the turbine, the steam gains kinetic energy by passing through a nozzle (a fixed nozzle in an impulse type turbine or the fixed blades in a reaction type turbine). When the steam leaves the nozzle it is moving at high velocity towards the blades of the turbine rotor. A force is created on the blades due to the pressure of the vapor on the blades causing them to move. A generator or other such device can be placed on the shaft, and the energy that was in the steam can now be stored and used. The steam leaves the turbine as a saturated vapor (or liquid-vapor mix depending on application) at a lower temperature and pressure than it entered with and is sent to the condenser to be cooled. The first law enables us to find an formula for the rate at which work is developed per unit mass. Assuming there is no heat transfer to the surrounding environment and that the changes in kinetic and potential energy are negligible compared to the change in specific enthalpy we arrive at the following equation

$$\frac{\dot{W}}{\dot{m}} = h_3 - h_4$$

where

- $\dot{W}$ is the rate at which work is developed per unit time
- $\dot{m}$ A□is the rate of mass flow through the turbine

Isentropic efficiency

To measure how well a turbine is performing we can look at its isentropic efficiency. This compares the actual performance of the turbine with the performance that would be achieved by an ideal, isentropic, turbine. When calculating this efficiency, heat lost to the surroundings is assumed to be zero. The starting pressure and temperature is the same for both the actual and the ideal turbines, but at turbine exit the energy content ('specific enthalpy') for the actual turbine is greater than that for the ideal turbine because of irreversibility in the actual turbine. The specific enthalpy is evaluated at the same pressure for the actual and ideal turbines in order to give a good comparison between the two.

The isentropic efficiency is found by dividing the actual work by the ideal work.

$$\eta_t = \frac{h_3 - h_4}{h_3 - h_{4s}}$$

where

- h_3 is the specific enthalpy at state three
- h_4 is the specific enthalpy at state 4 for the actual turbine
- h_{4s} is the specific enthalpy at state 4s for the isentropic turbine

(but note that the adjacent diagram does not show state 4s: it is vertically below state 3)

DIRECT DRIVE

A direct-drive 5 MW steam turbine fuelled with biomass

Electrical power stations use large steam turbines driving electric generators to produce most (about 80%) of the world's electricity. The advent of large steam turbines made central-station electricity generation practical, since reciprocating steam engines of large rating became very bulky, and operated at slow speeds. Most central stations are fossil fuel power plants and nuclear power plants; some installations use geothermal steam, or use concentrated solar power (CSP) to create the steam. Steam turbines can also be used directly to drive large centrifugal pumps, such as feedwater pumps at a thermal power plant.

The turbines used for electric power generation are most often directly coupled to their generators. As the generators must rotate at constant synchronous speeds according to the frequency of the electric power system, the most common speeds are 3,000 RPM for 50 Hz systems, and 3,600 RPM for 60 Hz systems. Since nuclear reactors have lower temperature limits than fossil-fired plants, with lower steam quality, the turbine generator sets may be arranged to operate at half these speeds, but with four-pole generators, to reduce erosion of turbine blades.

MARINE PROPULSION

Turbinia, 1894, the first steam turbine-powered ship

High and low pressure turbines forSS *Maui*.

Parsons turbine from the 1928 Polish destroyer *Wicher*.

In steamships, advantages of steam turbines over reciprocating engines are smaller size, lower maintenance, lighter weight, and lower vibration. A steam turbine is only efficient when operating in the thousands of RPM, while the most effective propeller designs are for speeds less than 300 RPM; consequently, precise (thus expensive) reduction gears are usually required, although numerous early ships through World War I, such as *Turbinia*, had direct drive from the steam turbines to the propeller shafts.

Another alternative is turbo-electric transmission, in which an electrical generator run by the high-speed turbine is used to run one or more slow-speed electric motors connected to the propeller shafts; precision gear cutting may be a production bottleneck during wartime. Turbo-electric drive was most used in large US warships designed during World War I and in some fast liners, and was used in some troop transports and mass-production destroyer escorts in World War II.

The higher cost of turbines and the associated gears or generator/motor sets is offset by lower maintenance requirements and the smaller size of a turbine when compared to a reciprocating engine having an equivalent power, although the fuel costs are higher than a diesel engine because steam turbines have lower thermal efficiency. To reduce fuel costs the thermal efficiency of both types of engine have been improved over the years. Today, propulsion steam turbine cycle efficiencies have yet to break 50%, yet diesel engines routinely exceed 50%, especially in marine applications. Diesel power plants also have lower operating costs since fewer operators are required. Thus, conventional steam power is used in very few new ships. An exception is LNG carriers which often find it more economical to use boil-off gas with a steam turbine than to re-liquify it.

Nuclear-powered ships and submarines use a nuclear reactor to create steam for turbines. Nuclear power is often chosen where diesel power would be impractical (as in submarine applications) or the logistics of refuelling pose significant problems (for example,icebreakers).

It has been estimated that the reactor fuel for the Royal Navy's *Vanguard*-class submarines is sufficient to last 40 circumnavigations of the globe – potentially sufficient for the vessel's entire service life. Nuclear propulsion has only been applied to a very few commercial vessels due to the expense of maintenance and the regulatory controls required on nuclear systems and fuel cycles.

Early development

The development of steam turbine marine propulsion from 1894-1935 was dominated by the need to reconcile the high efficient speed of the turbine with the low efficient speed (less than 300 rpm) of the ship's propeller at an overall cost competitive with reciprocating engines. In 1894, efficient reduction gears were not available for the high powers required by ships, so direct drive was necessary. In*Turbinia*, which has direct drive to each propeller shaft, the efficient speed of the turbine was reduced after initial trials by directing the steam flow through all three direct drive turbines (one on each shaft) in series, probably totaling around 200 turbine stages operating in series. Also, there were three propellers on each shaft for operation at high speeds. The high shaft speeds of the era are represented by one of the first US turbine-powered destroyers, USS *Smith*, launched in 1909, which had direct drive turbines and whose three shafts turned at 724 rpm at 28.35 knots. The use of turbines in several casings exhausting steam to each other in series became standard in most subsequent marine propulsion applications, and is a form of cross-compounding.

The first turbine was called the high pressure (HP) turbine, the last turbine was the low pressure (LP) turbine, and any turbine in between was an intermediate pressure (IP) turbine. A much later arrangement than *Turbinia* can be seen on RMS *Queen Mary* in Long Beach, California, launched in 1934, in which each shaft is powered by four turbines in series connected to the ends of the two input shafts of a single-reduction gearbox. They are the HP, 1st IP, 2nd IP, and LP turbines.

Cruising machinery and gearing

The quest for economy was even more important when cruising speeds were considered. Cruising speed is roughly 50% of a warship's maximum speed and 20-25% of its maximum power level. This would be a speed used on long voyages when fuel economy is desired. Although this brought the propeller speeds down to an efficient range, turbine efficiency was greatly reduced, and early turbine ships had poor cruising ranges. A solution that proved useful through most of the steam turbine propulsion era was the cruising turbine. This was an extra turbine to add even more stages, at first attached directly to one or more shafts, exhausting to a stage partway along the HP turbine, and not used at high speeds.

As reduction gears became available around 1911, some ships, notably the battleship USS *Nevada*, had them on cruising turbines while retaining direct drive main turbines. Reduction gears allowed turbines to operate in their efficient range at a much higher speed than the shaft, but were expensive to manufacture.

Cruising turbines competed at first with reciprocating engines for fuel economy. An example of the retention of reciprocating engines on fast ships was the famous RMS *Titanic*of 1911, which along with her sisters RMS *Olympic* and HMHS *Britannic* had triple-expansion engines on the two outboard shafts, both exhausting to an LP turbine on the center shaft. After adopting turbines with the *Delaware*-class battleships launched in 1909, the United States Navy reverted to reciprocating machinery on the *New York*-class battleships of 1912, then went back to turbines on *Nevada* in 1914. The lingering fondness for reciprocating machinery was because the US Navy had no plans for capital ships exceeding 21 knots until after World War I, so top speed was less important than economical cruising. The United States had acquired the Philippines and Hawaii as territories in 1898, and lacked the British Royal Navy's worldwide network of coaling stations. Thus, the US Navy in 1900-1940 had the greatest need of any nation for fuel economy, especially as the prospect of war with Japan arose following World War I. This need was compounded by the US not launching any cruisers 1908-1920, so destroyers were required to perform long-range missions usually assigned to cruisers. So, various cruising solutions were fitted on US destroyers launched 1908-1916. These included small reciprocating engines and geared or ungeared cruising turbines on one or two shafts. However, once fully geared turbines proved economical in initial cost and fuel they were rapidly adopted, with cruising turbines also included on most ships. Beginning in 1915 all new Royal Navy destroyers had fully geared turbines, and the United States followed in 1917.

In the Royal Navy, speed was a priority until the Battle of Jutland in mid-1916 showed that in the battlecruisers too much armour had been sacrificed in its pursuit. The British used exclusively turbine-powered warships from 1906. Because they recognized that a significant cruising range would be desirable given their world-wide empire, some warships, notably the *Queen Elizabeth*-class battleships, were fitted with cruising turbines from 1912 onwards following earlier experimental installations.

In the US Navy, the *Mahan*-class destroyers, launched 1935-36, introduced double-reduction gearing. This further increased the turbine speed above the shaft speed, allowing smaller turbines than single-reduction gearing. Steam pressures and temperatures were also increasing progressively, from 300 psi/425 F (2.07 MPa/218 C)(saturation temperature) on the World War I-era *Wickes* class to 615 psi/850 F (4.25 MPa/454 C) superheated steam on some World War II *Fletcher*-class destroyers and later ships.A standard configuration emerged of an axial-flow high pressure turbine (sometimes with a cruising turbine attached) and a double-axial-flow low pressure turbine connected to a double-reduction gearbox.

This arrangement continued throughout the steam era in the US Navy and was also used in some Royal Navy designs. Machinery of this configuration can be seen on many preserved World War II-era warships in several countries. When US Navy warship construction resumed in the early 1950s, most surface combatants and aircraft carriers used 1,200 psi/950 F (8.28 MPa/510 C) steam. This continued until the end of the US Navy steam-powered warship era with the *Knox*-classfrigates of the early 1970s. Amphibious and auxiliary ships continued to use 600 psi (4.14 MPa) steam post-World War II, with USS *Iwo Jima*, launched in 2001, possibly the last non-nuclear steam-powered ship built for the US Navy.

Turbo-electric drive

NS *50 Let Pobedy*, a nuclear icebreaker with nuclear-turbo-electric propulsion

Turbo-electric drive was introduced on the battleship USS *New Mexico*, launched in 1917. Over the next eight years the US Navy launched five additional turbo-electric-powered battleships and two aircraft carriers (initially ordered as *Lexington*-class battlecruisers). Ten more turbo-electric capital ships were planned, but cancelled due to the limits imposed by the Washington Naval Treaty. Although*New Mexico* was refitted with geared turbines in a 1931-33 refit, the remaining turbo-electric ships retained the system throughout their careers. This system used two large steam turbine generators to drive an electric motor on each of four shafts. The system was less costly initially than reduction gears and made the ships more maneuverable in port, with the shafts able to reverse rapidly and deliver more reverse power than with most geared systems. Some ocean liners were also built with turbo-electric drive, as were some troop transports and mass-production destroyer escorts in World War II. However, when the US designed the "treaty cruisers", beginning withUSS *Pensacola* launched in 1927, geared turbines were used for all fast steam-powered ships thereafter.

Current usage

Since the 1980s, steam turbines have been replaced by gas turbines on fast ships and by diesel engines on other ships; exceptions are nuclear-powered ships and submarinesand LNG carriers. In the U.S. Navy, the conventionally powered steam turbine is still in use on all but one of the Wasp-class amphibious assault ships. The U.S. Navy also operates steam turbines on their nuclear powered Nimitz-class and Ford-class aircraft carriers along with all of their nuclear submarines (Ohio-,Los Angeles-, Seawolf-, andVirginia-classes). The Royal Navy decommissioned its last conventional steam-powered *Leander*-class frigate in 1993, also converting its sole Type 82 destroyer, HMS *Bristol*, into a training ship that same year. In 2013, the French Navy ended its steam era with the decommissioning of its last *Tourville*-class frigate.

Of the remaining blue-water navies, the Russian and Chinese navies currently operate steam-powered *Kuznetsov*-class aircraft carriers and

Sovremenny-class destroyers, with China also operating steam-powered *Luda*-class destroyers. The Indian Navy currently operates two conventional steam-powered carriers, INS *Viraat*, a former British *Centaur*-class aircraft carrier (to be decommissioned in 2016), and INS *Vikramaditya*, a modified *Kiev*-class aircraft carrier; it also operates three *Brahmaputra*-class frigatescommissioned in the early 2000s and two *Godavari*-class frigates currently in the process of being decommissioned. The JDS *Kurama*, the last steam-powered JMSDF *Shirane*-class destroyer, will be decommissioned and replaced in 2017. Most other naval forces either retired or re-engined their steam-powered warships by 2010; as of 2015, the Brazilian Navy operates *São Paulo*, a former French *Clemenceau*-class aircraft carrier, while the Mexican Navy currently operates four former U.S. *Knox*-class frigates and two former U.S. *Bronstein*-class frigates. The Peruvian Navy currently operates the former Dutch *De Zeven Provinciën*-class cruiser *BAP Almirante Grau*; the Ecuadorian Navycurrently operates two modified *Leander*-class frigates.

LOCOMOTIVES

A steam turbine locomotive engine is a steam locomotive driven by a steam turbine.

The main advantages of a steam turbine locomotive are better rotational balance and reduced hammer blow on the track. However, a disadvantage is less flexible output power so that turbine locomotives were best suited for long-haul operations at a constant output power.

The first steam turbine rail locomotive was built in 1908 for the Officine Meccaniche Miani Silvestri Grodona Comi, Milan, Italy. In 1924 Krupp built the steam turbine locomotive T18 001, operational in 1929, for Deutsche Reichsbahn.

TESTING

British, German, other national and international test codes are used to standardize the procedures and definitions used to test steam turbines. Selection of the test code to be used is an agreement between the purchaser and the manufacturer, and has some significance to the design of the turbine and associated systems. In the United States, ASMEhas produced several performance test codes on steam turbines. These include ASME PTC 6-2004, Steam Turbines, ASME PTC 6.2-2011, Steam Turbines in Combined Cycles, PTC 6S-1988, Procedures for Routine Performance Test of Steam Turbines. These ASME performance test codes have gained international recognition and acceptance for testing steam turbines. The single most important and differentiating characteristic of ASME performance test codes, including PTC 6, is that the test uncertainty of the measurement indicates the quality of the test and is not to be used as a commercial tolerance.

MARINE STEAM ENGINE

A marine steam engine is a steam engine that is used to power a ship or boat. The marine steam engines of the reciprocating type, which were in use from the inception of the steamboat in the early 19th century to their last years of large-scale manufacture during World War II. Reciprocating steam engines were progressively replaced in marine applications during the 20th century by steam turbines and marine diesel engines.

HISTORY

The first commercially successful steam engine was developed by Thomas Newcomen in 1712. The steam engine improvements brought forth by James Watt in the later half of the 18th century greatly improved steam engine efficiency and allowed more compact engine arrangements. Successful adaptation of the steam engine to marine applications in England would have to wait until almost a century later after Newcomen, when Scottish engineer William Symington built the world's "first practical steamboat", the *Charlotte Dundas*, in 1802. In 1807, the American Robert Fulton built the world's first commercially successful steamboat, simply known as the *North River Steamboat*, and powered by a Watt engine.

Following Fulton's success, steamboat technology developed rapidly on both sides of the Atlantic. Steamboats initially had a short range and were not particularly seaworthy due to their weight, lack of horsepower, and tendency to break down, but they were employed successfully along rivers and canals, and for short journeys along the coast. The first successful transatlantic crossing by a steamship occurred in 1819 when *Savannah* sailed from Savannah, Georgia to Liverpool, England. The first steamship to make regular transatlantic crossings was the sidewheel steamer *Great Western* in 1838.

As the 19th century progressed, marine steam engine and steamship technology developed alongside it. Paddle propulsion gradually gave way to the screw propeller, and the introduction of iron and later steel hulls to replace the traditional wooden hull allowed ships to grow ever larger, necessitating steam power plants that were increasingly complex and powerful.

TYPES OF MARINE STEAM ENGINE

A wide variety of reciprocating marine steam engines were developed over the course of the 19th century. The two main methods of classifying such engines are by *connection mechanism* and *cylinder technology*.

Most early marine engines had the same cylinder technology (simple expansion, see below) but a number of different methods of supplying power to the crankshaft (i.e. connection mechanism) were in use. Thus, early marine engines are classified mostly according to their connection mechanism. Some common connection mechanisms were side-lever, steeple, walking beam and direct-acting. However, steam engines can also be classified according to their cylinder technology (simple-expansion, compound, annular etc.). One can therefore sometimes find examples of engines which were classified under both methods, such as the compound walking beam (*compound* being the cylinder technology and *walking beam* being the connection method). Over time, as most engines became direct-acting but cylinder technologies were growing more complex, engines began to be classified solely according to their cylinder technology instead.

ENGINES CLASSIFIED BY CONNECTION MECHANISM

Side-lever

The side-lever engine was the first type of steam engine to be widely adopted for marine use in Europe. In the early years of steam navigation (from c1815), the side-lever was the most common type of marine engine for inland waterway and coastal service in Europe, and it remained for many years the preferred engine for oceangoing service on both sides of the Atlantic.

The side-lever was an adaptation of the earliest form of steam engine, the beam engine. The typical side-lever engine had a pair of heavy horizontal iron beams, known as side-levers, each secured in the centre by a pin near the base of the engine, allowing the levers to pivot through a limited arc. The engine cylinder stood vertically between this pair of levers at one end, with the piston rod attached to a horizontal crosshead, from each end of which a vertical rod, known as a side-rod, extended down each side of the cylinder to connect to the end of the side-lever on the same side. The far ends of the two side-levers were connected to one another by a horizontal crosstail, from which extended a single, common connecting rod which operated the crankshaft as the levers rocked up and down around the central pin.

The main disadvantages of the side-lever engine were that it was large and heavy, and for inland waterway and coastal service, it was soon replaced by lighter and more efficient designs. It remained the dominant engine type for oceangoing service through much of the first half of the 19th century however, due to its relatively low centre of gravity which gave ships more stability in heavy seas. It was also a common early engine type for warships, since its relatively low height made it less susceptible to battle damage. From the first Royal Navy steam vessel in 1820 until 1840, 70 steam vessels entered service, the majority with side-lever engines, using boilers set to 4psi maximum pressure. The low steam pressures dictated the large cylinder sizes for the side-lever engines, though the effective pressure on the piston was the difference between the boiler pressure and the vacuum in the condenser.

The side-lever engine was a paddlewheel engine and was not suitable for driving screw propellers. The last ship built for transatlantic service to be fitted with a side-lever engine was the Cunard Line's paddle steamer RMS *Scotia*, considered an anachronism when it entered service in 1862.

Grasshopper

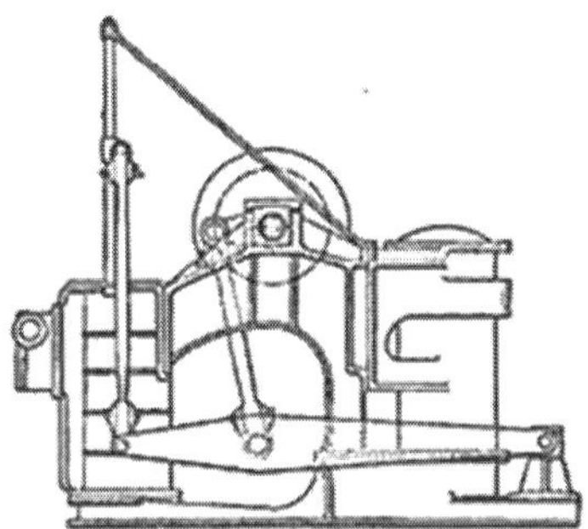

Diagram of a grasshopper engine

The grasshopper or 'half-lever' engine was a variant of the side-lever engine. The grasshopper engine differs from the conventional side-lever in that the location of the lever pivot and connecting rod are more or less reversed, with the pivot located at one end of the lever instead of the centre, while the connecting rod is attached to the lever between the cylinder at one end and the pivot at the other.

Chief advantages of the grasshopper engine were cheapness of construction and robustness, with the type said to require less maintenance than any other type of marine steam engine. Another advantage is that the engine could be easily started from any crank position. Like the conventional side-lever engine however, grasshopper engines were disadvantaged by their weight and size. They were mainly used in small watercraft such as riverboats and tugs.

Crosshead (square)

The crosshead engine, also known as a square, sawmill or A-frame engine, was a type of paddlewheel engine used in the United States. It was the most common type of engine in the early years of American steam navigation.

The crosshead engine is described as having a vertical cylinder above the crankshaft, with the piston rod secured to a horizontal crosshead, from each end of which, on opposite sides of the cylinder, extended a connecting rod which rotated its own separate crankshaft. The crosshead operated within vertical guides that enabled the assembly to maintain the correct path as it moved. The engine's alternative name "A-frame" presumably derived from the shape of the frames supporting these guides. Some crosshead engines had more than one cylinder, in which case the piston rods were usually all connected to the same crosshead. An unusual feature of early examples of this type of engine was the installation of flywheels—geared to the crankshafts—which were thought necessary to ensure smooth operation. These gears could apparently be very noisy in operation. Because the cylinder was placed above the crankshaft in this type of engine, it had a high center of gravity and was therefore deemed unsuitable for oceangoing service, so that its use was largely confined to vessels plying inland waterways. As marine engines grew steadily larger and heavier through the course of the century, the high center of gravity of square crosshead engines became increasingly impractical, leading to their abandonment by the 1840s in favor of the walking beam engine.

The name of this engine can sometimes lead to confusion as "crosshead" is also an alternative name for the steeple engine. Many sources thus prefer to refer to it by its informal name of "square" engine to avoid confusion. Additionally, the marine crosshead or square engine described in this section should not be confused with the term "square engine" as applied to internal combustion engines, which in the latter case refers to an engine whose bore is equal to its stroke.

Model of a crosshead or "square" engine, showing location of engine cylinder above the crankshaft; also piston rod, crosshead, connecting rods and paddlewheels

The 1836 paddle steamer *New York*. Between the paddlewheels is the tall square or "A-frame" engine, within which can be seen the long piston rod, near the top of its stroke, making a "T" with the horizontal crosshead

Walking beam

The walking beam, also known as a "vertical beam", "overhead beam", or simply "beam", was another early adaptation of the beam engine, but its use was confined almost entirely to the United States. After its introduction, the walking beam quickly became the most popular engine type in America for inland waterway and coastal service, and the type proved to have remarkable longevity, with walking beam engines still being occasionally manufactured as late as the 1940s. In marine applications, the beam itself was generally reinforced with iron struts that gave it a characteristic diamond shape, although the supports on which the beam rested were often built of wood. The adjective "walking" was applied because the beam, which rose high above the ship's deck, could be seen operating, and its rocking motion was (somewhat fancifully) likened to a walking motion.

Walking beam engines were a type of paddlewheel engine and were rarely used for powering propellers. They were used primarily for ships and boats working in rivers, lakes and along the coastline, but were a less popular choice for seagoing vessels because the great height of the engine made the vessel less stable in heavy seas. They were also of limited use militarily, because the engine was exposed to enemy fire and could thus be easily disabled. Their popularity in the United States was due primarily to the fact that the walking beam engine was well suited for the shallow-draft boats which operated in America's shallow coastal and inland waterways.

Walking beam engines remained popular with American shipping lines and excursion operations right into the early 20th century. Although the walking beam engine was technically obsolete in the later 19th century, it remained popular with excursion steamer passengers who expected to see the "walking beam" in motion. There were also technical reasons for retaining the walking beam engine in America, as it was easier to build, requiring less precision in its construction. Wood could be used for the main frame of the engine, at a much lower cost than typical practice of using iron castings for more modern engine designs. Fuel was also much cheaper in America than in Europe, so the lower efficiency of the walking beam engine was less of a consideration. The Philadelphia shipbuilder Charles H. Cramp blamed America's general lack of competitiveness with the British shipbuilding industry in the mid-to-late 19th century upon the conservatism of American domestic shipbuilders and shipping line owners, who doggedly clung to outdated technologies like the walking beam and its associated paddlewheel long after they had been abandoned in other parts of the world.

USS *Delaware* (1861). The vessel's diamond shaped "walking beam" can clearly be seen amidships

Steeple

The steeple engine, sometimes referred to as a "crosshead" engine, was an early attempt to break away from the beam concept common to both the walking beam and side-lever types, and come up with a smaller, lighter, more efficient design. In a steeple engine, the vertical oscillation of the piston is not converted to a horizontal rocking motion as in a beam engine, but is instead used to move an assembly, composed of a crosshead and two rods, through a vertical guide at the top of the engine which in turn rotates the crankshaft connecting rod below.

In early examples of the type, the crosshead assembly was rectangular in shape, but over time it was refined into an elongated triangle. The triangular assembly above the engine cylinder gives the engine its characteristic "steeple" shape, hence the name.

Steeple engines were tall like walking beam engines, but much narrower laterally, saving both space and weight. Because of their height and high centre of gravity, they were, like walking beams, considered to be less appropriate for oceangoing service, but they remained highly popular for several decades, especially in Europe, for inland waterway and coastal vessels.

Steeple engines began to appear in steamships in the 1830s and the type was perfected in the early 1840s by the British shipbuilder David Napier. The steeple engine was gradually superseded by the various types of direct-acting engine.

Siamese

The Siamese engine, also referred to as the "double cylinder" or "twin cylinder" engine, was another early alternative to the beam or side-lever engine. This type of engine had two identical, vertical engine cylinders arranged side-by-side, whose piston rods were attached to a common, T-shaped crosshead.

The vertical arm of the crosshead extended down between the two cylinders and was attached at the bottom to both the crankshaft connecting rod and to a guide block that slid between the vertical sides of the cylinders, enabling the assembly to maintain the correct path as it moved.

The Siamese engine was invented by British engineer Joseph Maudslay (son of Henry), but although he invented it after his oscillating engine, it failed to achieve the same widespread acceptance, as it was only marginally smaller and lighter than the side-lever engines it was designed to replace. It was however used on a number of mid-century warships, including the first warship fitted with a screw propeller, HMS *Rattler*.

Direct acting

There are two definitions of a direct-acting engine encountered in 19th-century literature. The earlier definition applies the term "direct-acting" to any type of engine other than a beam (i.e. walking beam, side-lever or grasshopper) engine. The later definition only uses the term for engines which apply their power directly to the crankshaft via the piston rod and/or connecting rod. Unlike the side-lever or beam engine, a direct-acting engine could be readily adapted to power either paddlewheels or a propeller. As well as offering a lower profile, direct-acting engines had the advantage of being smaller and weighing considerably less than beam or side-lever engines. The Royal Navy found that on average a direct-acting engine (early definition) weighed 40% less and required an engine room only two thirds the size of that for a side-lever of equivalent power. One disadvantage of such engines is that they were more prone to wear and tear and thus required more maintenance.

Oscillating

An oscillating engine was a type of direct-acting engine that was designed to achieve further reductions in engine size and weight. Oscillating engines had the piston rods connected directly to the crankshaft, dispensing with the need for connecting rods. In order to achieve this aim, the engine cylinders were not immobile as in most engines, but secured in the middle by trunnions which allowed the cylinders themselves to pivot back and forth as the crankshaft rotated, hence the term *oscillating*. Steam was supplied and exhausted through the trunnions. The oscillating motion of the cylinder was usually used to line up ports in the trunnions to direct the steam feed and exhaust to the cylinder at the correct times. However, separate valves may be provided, controlled by the oscillating motion. This allows the timing to be varied to enable expansive working. (As, for example, the engine in the paddle ship PD Krippen.) This compromises the advantage of simplicity but still retains the advantage of compactness.

The first patented oscillating engine was built by Joseph Maudslay in 1827, but the type is considered to have been perfected by John Penn. Oscillating engines remained a popular type of marine engine for much of the 19th century.

Model of a Maudslay oscillating engine

Oscillating engine built in 1853 by J. & A. Blyth of London for the Austrian paddle steamer *Orsova*

Trunk

The trunk engine, another type of direct-acting engine, was originally developed as a means of reducing an engine's height while retaining a long stroke (a long stroke was considered important at this time because it reduced the strain on components).

A trunk engine locates the connecting rod *within* a large-diameter hollow piston rod. This "trunk" carries almost no load. The interior of the trunk is open to outside air, and is wide enough to accommodate the side-to-side motion of the connecting rod, which links a gudgeon pin at the piston head to an outside crankshaft.

The walls of the trunk were either bolted to the piston or cast as one piece with it, and moved back and forth with it. The working portion of the cylinder is annular or ring-shaped, with the trunk passing through the centre of the cylinder itself.

Early examples of trunk engines had vertical cylinders; however, it was quickly realized that the type was compact enough to be laid horizontally across the keel. In this configuration, it was very useful to navies, as it had a profile low enough to fit entirely below a ship's waterline, where it would be as safe as possible from enemy fire. The type was generally produced for military service by John Penn.

Trunk engines were quite common on mid-19th century warships, and were also to be found in commercial vessels, where though valued for their compact size and low centre of gravity, they proved expensive to operate. Trunk engines however proved poorly adapted to the higher boiler pressures that became prevalent in the latter half of the 19th century, and were abandoned in favour of other solutions.

Normally large engines, a small mass-produced, high-revolution, high-pressure version was produced for the Crimean War. In being quite effective, the type persisted in later gunboats. An original trunk engine of the gunboat type exists in the Western Australian Museum in Fremantle. After sinking in

1872, it was raised in 1985 from the SS *Xantho*and can now be turned over by hand. The engine's mode of operation, illustrating its compact nature, could be viewed on the *Xantho* project's website.

Trunk engine illustration

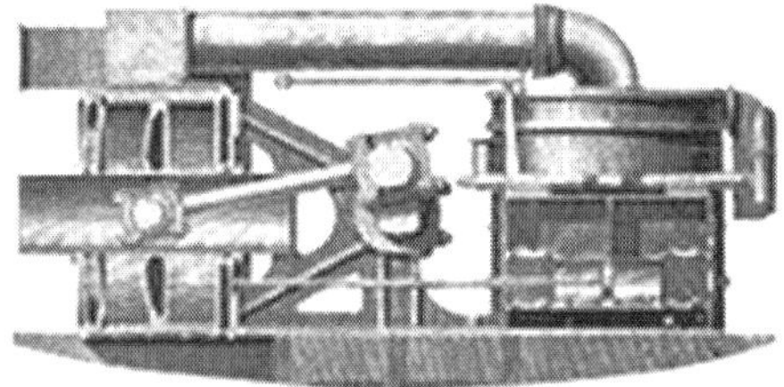

Cutaway view of trunk engine ofHMS *Bellerophon*, showing (on the left) engine cylinder, annular piston and trunk assembly, and connecting rod inside trunk

Looking down at the trunk engine ofHMS *Warrior* (1860). The connecting rod can be seen emerging from the trunk at right.

Vibrating lever

The vibrating lever, or half-trunk engine, was a development of the conventional trunk engine conceived by Swedish-American engineerJohn Ericsson. Ericsson needed a small, low-profile engine like the trunk engine to power the U.S. Federal government's monitors, a type of warship developed during the American Civil War that had very little space for a conventional powerplant. The trunk engine itself was however unsuitable for this purpose because the preponderance of weight was on the side of the engine containing the cylinder and trunk, a problem which could not be compensated for on the small monitor warships.

Ericsson resolved this problem by placing two horizontal cylinders back-to-back in the middle of the engine, working two "vibrating levers", one on each side, which by means of shafts and additional levers rotated a centrally located crankshaft. Vibrating lever engines were later used in some other warships and merchant vessels, but their use was confined to ships built in the United States and in Ericsson's native country of Sweden, and as they had few advantages over more conventional engines, were soon supplanted by other types.

Back acting

The back-acting engine, also known as the return connecting rod engine, was another engine designed to have a very low profile. The back-acting engine was in effect a modified steeple engine, laid horizontally across the keel of a ship rather than standing vertically above it. Instead of the triangular crosshead assembly found in a typical steeple engine however, the back-acting engine generally utilized a set of two or more elongated, parallel piston rods terminating in a crosshead to perform the same function. The term "back-acting" or "return connecting rod" derives from the fact that the connecting rod "returns" or comes back from the side of the engine opposite the engine cylinder to rotate a centrally located crankshaft.

Back-acting engines were another type of engine popular in both warships and commercial vessels in the mid-19th century, but like many other engine types in this era of rapidly changing technology, they were eventually abandoned for other solutions. There is only one back-acting engine known to be still in existence—that of the TV *Emery Rice*(formerly USS *Ranger*), now the centerpiece of a display at the American Merchant Marine Museum.

Vertical

As steamships grew steadily in size and tonnage through the course of the 19th century, the need for low profile, low centre-of-gravity engines correspondingly declined. Freed increasingly from these design constraints, engineers were able to revert to simpler, more efficient and more easily maintained designs. The result was the growing dominance of the so-called "vertical" engine (more correctly known as the vertical inverted direct acting engine).

In this type of engine, the cylinders are located directly above the crankshaft, with the piston rod/connecting rod assemblies forming a more or less straight line between the two. The configuration is similar to that of a modern internal combustion engine (one notable difference being that the steam engine is double acting, whereas almost all internal combustion engines generate power only in the downward stroke). Vertical engines are sometimes referred to as "hammer", "forge hammer" or "steam hammer" engines, due to their roughly similar appearance to another common 19th-century steam technology, the steam hammer.

Vertical engines came to supersede almost every other type of marine steam engine toward the close of the 19th century. Because they became so common, vertical engines are not usually referred to as such, but are instead referred to based upon their cylinder technology, i.e. as compound, triple-expansion, quadruple-expansion etc. It should be noted that the term "vertical" for this type of engine is imprecise, since technically any type of steam engine is "vertical" if the cylinder is vertically oriented. An engine described as "vertical" should therefore not be assumed to be of the vertical inverted direct-acting type unless the term "vertical" is unqualified.

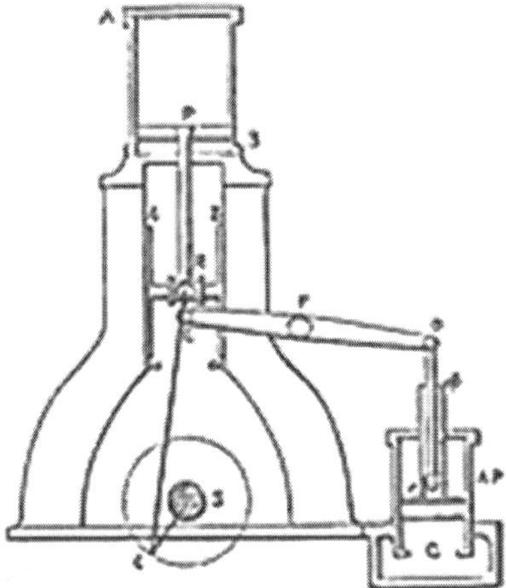

Diagram of a simple "hammer" engine

Vertical triple-expansion engine ofUSS *Wisconsin* (BB-9). The typical vertical engine arrangement of cylinder, piston rod, connecting rod and crankshaft can clearly be seen in this photo.

ENGINES CLASSIFIED BY CYLINDER TECHNOLOGY

SIMPLE EXPANSION

A simple-expansion engine is a steam engine that expands the steam through only one stage, which is to say, all its cylinders are operated at the same pressure. Since this was by far the most common type of engine in the early period of marine engine development, the term "simple expansion" is rarely encountered; rather, an engine is assumed to be simple-expansion unless otherwise stated.

Compound

A compound engine is a steam engine which operates cylinders through more than one stage, i.e., at different pressure levels. Compound engines were a method of improving efficiency. Up until the development of compound engines, steam engines used the steam only once before being recycled back to the boiler, but a compound engine recycles the steam into one or more larger, lower-pressure second cylinders first, in order to utilize more of its heat energy. Compound engines could be configured to either increase a ship's economy or its speed. Although broadly speaking a compound engine can refer to a steam engine with any number of different-pressure cylinders, the term usually refers to engines which expand steam through only two stages, i.e. those which operate cylinders at only two different pressures (or "double-expansion" engines).

Note that a compound engine can have more than one *set* of variable-pressure cylinders. For example, an engine might have two cylinders operating at pressure x and two operating at pressure y, or one cylinder operating at pressure x and three operating at pressure y. What makes it compound (or double-expansion) as opposed to multiple-expansion is that there are only two *pressures*, x and y. The first compound engine believed to have been installed in a ship was that fitted to *Henry Eckford* by the American engineer James P. Allaire in 1824. However, many sources attribute the "invention" of the marine compound engine to Glasgow's John Elder in the 1850s. Elder made improvements to the compound engine that made it safe and economical for ocean-crossing voyages for the first time.

Triple or multiple expansion

A triple-expansion engine is a compound engine that expands the steam in three stages, i.e. an engine which has cylinders operating at three different pressures. A quadruple-expansion engine expands the steam in four stages, and so on. The first successful commercial use was an engine built at Govan in Scotland by Alexander C. Kirk for the SS*Aberdeen* in 1881.

Multiple-expansion engine manufacture continued well into the 20th century. All 2,700 Liberty ships built by the United States during World War II were powered by triple-expansion engines, because the capacity of the US to manufacture marine steam turbines was entirely directed to the building of warships. The biggest manufacturer of triple-expansion engines during the war was the Joshua Hendy Iron Works. Toward the end of the war, turbine-powered Victory ships were manufactured in increasing numbers.

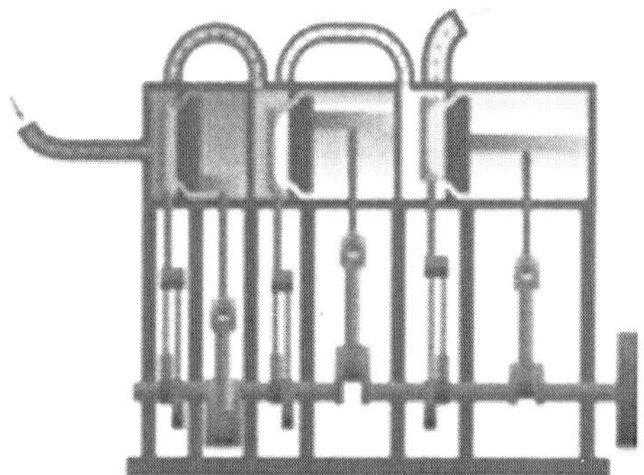

Animation of a typical vertical triple-expansion engine

A Joshua Hendy triple-expansion engine

A triple-expansion engine on theLydia Eva (steam drifter)

A triple-expansion engine on the 1907 oceangoing tug *Hercules*

140-ton – also described as 135-ton – vertical triple-expansion engine of the type used to powerWorld War II Liberty ships, assembled for testing prior to delivery. The engine is 21 feet (6.4 meters) long and 19 feet (5.8 meters) tall and was designed to operate at 76 rpmand propel a Liberty ship at about 11 knots (12.7 mph; 20.4 km/h).

Annular

An annular engine is an unusual type of engine that has an annular (ring-shaped) cylinder. Some of American pioneering engineer James P. Allaire's early compound engines were of the annular type, with a smaller, high-pressure cylinder placed in the centre of a larger, ring-shaped low-pressure cylinder. Trunk engines were another type of annular engine. A third type of annular marine engine which was sometimes produced utilized the Siamese engine connecting mechanism, but instead of two separate cylinders, had a single annular-shaped cylinder wrapped around the vertical arm of the crosshead.

OTHER TERMS

Some other terms are encountered in marine engine literature of the period. These terms, listed below, are usually used in conjunction with one or more of the basic engine classification terms listed above.

Simple

A simple engine is an engine with only one cylinder. Up until about the mid-19th century, most ships had engines with only one cylinder (although some vessels had more than one engine). Simple engines are always also simple-expansion engines by necessity.

Double acting

A double acting engine is an engine where steam is applied to both the up and down stroke of the piston. Earlier steam engines applied steam in only one direction, allowing momentum or gravity to return the piston to its starting place, but a double acting engine uses steam to force the piston in both directions, thus increasing RPM and power.Like the term "simple expansion", the term "double acting" is infrequently encountered in the literature since almost all marine engines were of the double acting type.

Vertical, horizontal, inclined, inverted

These terms refer to the orientation of the engine cylinder. A vertical cylinder stands vertically with its piston rod operating above it. An inverted cylinder (or "vertical inverted" cylinder) can be thought of as a vertical cylinder positioned upside down. With an inclined or horizontal type, the cylinder and piston are positioned at an incline or horizontally. An inclined inverted cylinder is an inverted cylinder operating at an incline. These terms are all generally used in conjunction with the engine types above. Thus, one may have a horizontal direct-acting engine, or an inverted walking beam, and so on.

Inclined and horizontal cylinders could be very useful in naval vessels as their orientation kept the engine profile as low as possible and thus less susceptible to damage. They could also be used in a low profile ship or to keep a ship's centre of gravity lower. In addition, inclined or horizontal cylinders had the advantage of reducing the amount of vibration by comparison with a vertical cylinder.

Geared

A geared engine or "geared screw" turns the propeller at a different rate to the engine's RPM. Early marine propeller engines were geared upward, which is to say the propeller was geared to run at a higher RPM than the engine itself. As engines became faster and more powerful through the latter part of the 19th century, gearing was often dispensed with and the propeller ran at the same RPM as the engine.

STEAM PLANTS

Water in the form of steam has the ability to store great amounts of energy. With it's ease of control and delivery, steam brought the advent of power to the shipping world. There are still some steam powered vessels such as ULCC (Ultra Large Crude Carrier) where steam turbines can provide the necessary, high power shaft requirements to propel the ship. However it's time as passed, most ships nowadays use the more economical diesel burning heavy fuels.

Although boilers may no longer be commonplace for ship propulsion they are almost guaranteed to be one boiler for various duties on board a ship. Duties like heating cargo, fuel, and accommodations. Some ships also use boilers for auxiliary power. Such as deck winches and pumps, where electrical machines would prove to be a hazard as in the oil industry.

STEAM THEORY

Within the boiler, fuel and air are force into the furnace by the burner. There, it burns to produce heat. From there, the heat (flue gases) travel throughout the boiler. The water absorbs the heat, and eventually absorb enough to change into a gaseous state - steam.

To the left is the basic theoretical design of a modern boiler. Boiler makers have developed various designs to squeeze the most energy out of fuel and to maximized its transfer to the water. But it all boils down, pardon the pun, to the basic design shown here. Below are a description of the most accepted variations of the basic principles (above left).

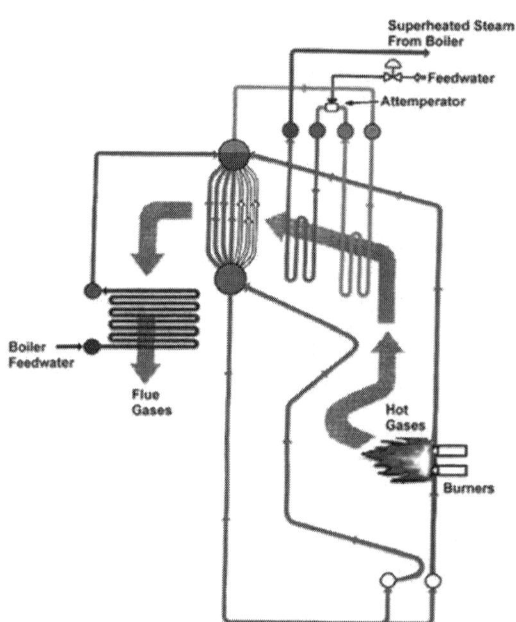

THE WATER TUBE BOILER

As you can see, the Babcock Marine Water Tube Boiler (below) looks very complicated. Thousands of tubes are placed in strategic location to optimize the exchange of energy from the heat to the water in the tubes. These types of boilers are most common because of their ability to deliver large quantities of steam.

The large tube like structure at the top of the boiler is called the steam drum. You could call it the heart of the boiler. That's where the steam collects before being discharged from the boiler. The hundreds of tube start and eventually end up at the steam drum.

Water enters the boiler, preheated, at the top. The hot water naturally circulates through the tubes down to the lower area where it is hot. The water heats up and flows back to the steam drum where the steam collects. Not all the water gets turn to steam, so the process starts again. Water keeps on circulating until it becomes steam.

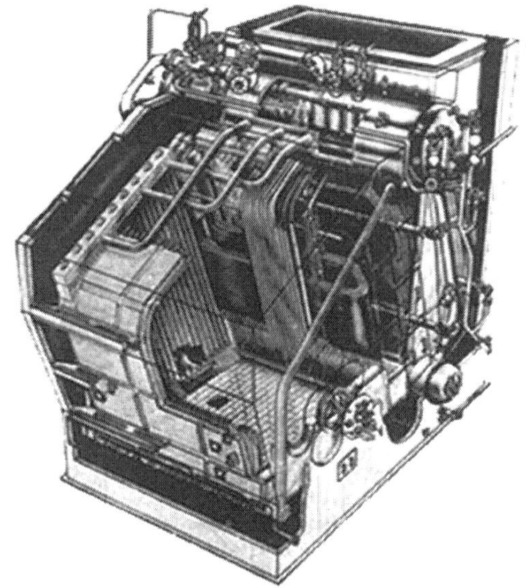

Meanwhile, the control system is taking the temperature of the steam drum, along with numerous other readings, to determine if it should keep the burner burning, or shut it down.

As well, sensors control the amount of water entering the boiler, this water is know as feed water. Feed water is not your regular drinking water. It is treated with chemicals to neutralize various minerals in the water, which untreated, would cling to the tubes clogging or worst, rusting them. This would make the boiler expensive to operate because it would not be very efficient.

On the fire side of the boiler, carbon deposit resulting from improper combustion or impurities in the fuel can accumulate on the outer surface of the water tube. This creates an insulation which quickly decrease the energy transfer from the heat to the water. To remedy this problem the engineer will carry out soot blowing. At a specified time the engineer uses a long tool and insert it into the fire side of the boiler. This device, which looks like a lance, has a tip at the end which "blows" steam. This blowing action of the steam "scrubs" the outside of the water tubes, cleaning the carbon build up.

Water tube boilers can have pressures from 7 bar (one bar = ~15 psi) to as high as 250 bar. The steam temperature's can vary between saturated steam, 100 degrees Celsius steam with particle of water, or be as high as 600 - 650 degrees Celsius, know as superheated steam or dry steam (all water particle have been turn to a gaseous state).

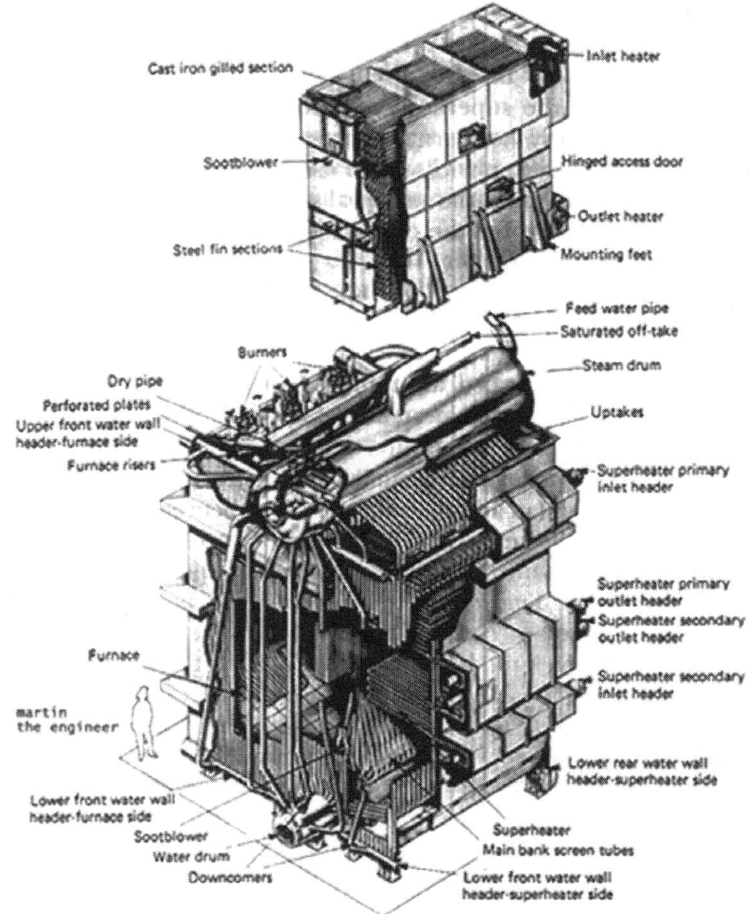

The performance of boiler is generally referred to as tons of steam produced in one hour. In water tube boilers that could be as low as 1.5 t/hr to as high as 2500 t/hr. The larger boilers would be land based, your local power company would most likely operate one. In British Columbia, large boilers are most common at Pulp and Paper plants.

Foster Wheeler (USA/UK), Babcock (USA/UK/Ger), Combustion Engineering (USA), and Kawasaki Heavy Industries (Japan) are some of the more prominent manufacturer of boilers. Click on the picture to the right to view a full size diagram of a Foster Wheeler ESD III water tube boiler.

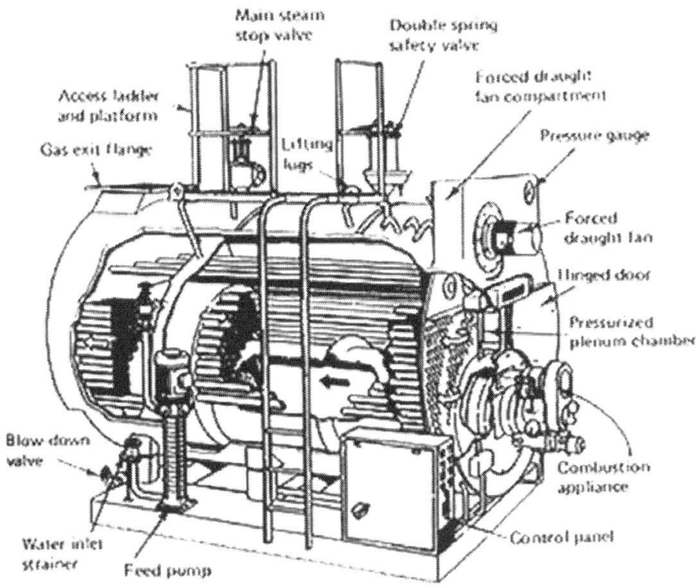

THE FIRE TUBE BOILER

This type of boilers started it all. This is the original design of boiler which brought the tide of power to the marine world. If you are ever in Vancouver, BC, the SS Master, a turn of the century tugboat, is open for the public to view at the Vancouver Maritime Museum. It is operational, and a fine example of ship using a fire tube boiler.

On a modern ship, the fire tube boiler meet the ship's heating needs and is generally not used for deck machinery. The steam produced will circulate through coils in the cargo tanks, fuel tanks, and accommodation heating system. They are generally supplied as a complete package, such as the one pictured above.

This is a single furnace, three pass type fire tube boiler. Heat - flue gases - travels through three different sets of tubes. All the tubes are surrounded by water which absorbs the heat. As the water turns to steam, pressure builds up within the boiler, once enough pressure has built up the engineer will open main steam outlet valve slowly, supplying steam for service. Fire tube boilers are also known as "smoke tube" and "donkey boiler".

... AND THE AUXILIARY BOILER

On smaller ships the auxiliary boiler can be a stand alone unit and would most likely be of the fire tube boiler arrangement as described above. But on a larger vessel it is more efficient for the auxiliary boiler to take advantage of the main engine's flue gases to heat the water. Basically this means that the hot gases from the main engine must pass through a heat exchanger (the auxiliary fire tube boiler) before exiting to the atmosphere.

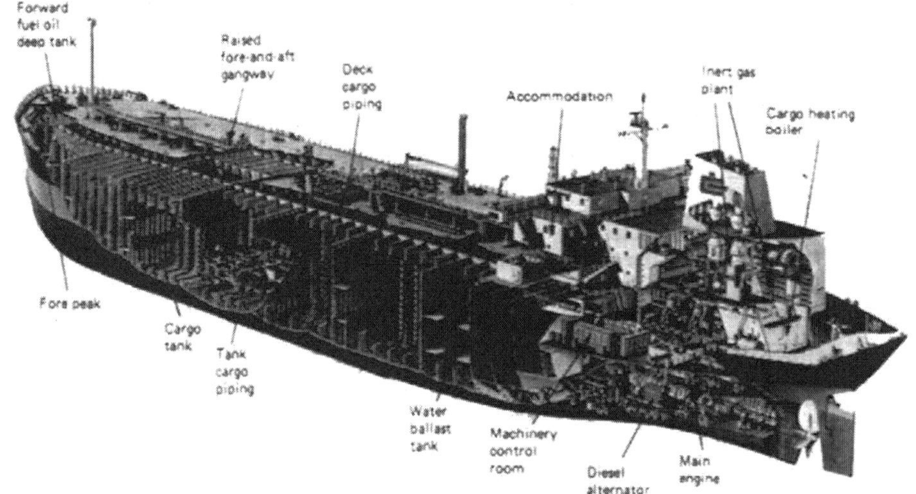

On this diagram, look for it above, and just aft of the main engine, near the exhaust stake of the ship. It is called the "cargo heating boiler".

As you can imagine if the ship's main engine was not running, there would be no hot flue gases to make steam. The auxiliary boiler also has a burner assembly which can be operated while the ship is in port or when the flue gases are not hot enough to provide the necessary steam.

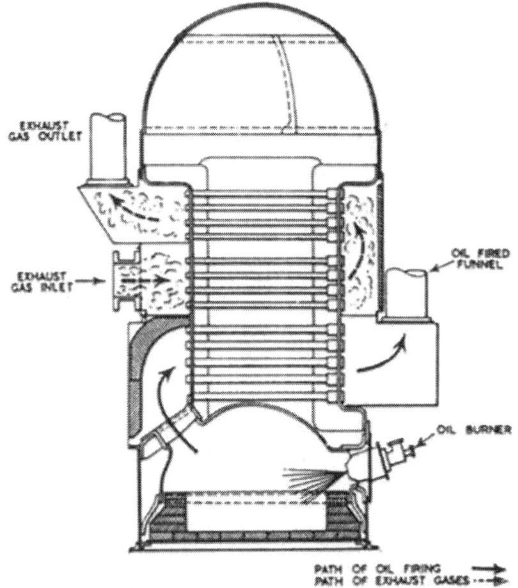

With this Cochran type boiler, the flow of flue gascs from the engine are controlled by a damper. Should the damper not allow engine flue gases through, the burner would automatically come on and provide heat for the water to absorb. It would do so until the controls of the damper allowed the flue gases to flow through the boiler providing the necessary heat for the water, the burner would then shut down.

USING THE STEAM TO MAKE THE SHIP GO !

Rotating the propeller is the ultimate goal of any power plant. As you have probably noticed, from the text and pictures above, there is no shaft. Which leads to the question:

"now that you have all this super energized steam, how do you get work from it ? "

A boilers is only one part of a larger operation, granted, it's a large part but most important part of the operation is it's ability to apply all this steam power.

THE RECIPROCATING STEAM ENGINE.

THEORY

- Every action has an equal and opposite reaction.
- Everything in nature reaches a balance.
- High pressure steam, 20bar, wants to 'go back' to being like everything else on the planet, which is water at 1 bar.

If you understand the above, basic, principle, engines become very logical machines.

In the earlier days the primary engine to transform the steam's heat energy to mechanical energy was done using a piston within a sealed housing. Valves in the sealed housing would allow steam to enter into the chamber, the steam restricted by the sealed housing would push on the piston, forcing it down.

This downward motion of the piston was transmitted to the crankshaft by a connecting rod. The illustration below is the best way to view the basic principle of the piston action.

This illustration, courtesy of Rick Boggs' Merchant Marine and Maritime Pages, is of a Triple Expansion Steam Engine. This type of design was very common at the earlier part of the 20th century. The SS Master, a tugboat on display at the Vancouver Maritime Museum, has a good example of a working triple expansion steam engine. The Famous RMS Titanic had two similar engines, except the Titanic's had an additional stage. They were known as quadruple expansion engine and operated on the same principle.

Although the model rotates a little fast, it clearly illustrates the action of the steam. The superheated steam (steam @ 101+degrees Celsius) will be used to "push" up or "down" three times in this engine.

The first time, where the steam has the most energy, the valve allows it to enter the small cylinder, on the topside of the piston. The expansion (pressure) of the steam pushes down on the area of the piston, rotating the crankshaft. The steam is then release by ports, near the end of it's stroke. The steam is then directed to the following cylinder. Here for a second time, by way of a valve, the steam enters the medium size cylinder and exert it's pressure on the area of the piston forcing it down. Finally, with most of the energy already spent, the steam enters the third and final stage of the engine as it did in the two previous stages. The steam enters the large diameter cylinder, pushes down the piston and exits the engine. The steam is then collected in a vacuum environment called a condenser, where the remaining heat in the steam is dispelled and changes state, back to being water. The water is then fed, or should I say recycled, as feedwater for the boiler.

The pistons of this engine are called double acting, which means that, not only does the piston get "pushed down" but it also gets "pushed up". The three stages describe above are also, simultaneously, happening to the underside of the piston. So steam enters the top of the piston, pushes it down, then the valve allows steam to enter the bottom of the piston, pushing it up.

THE STEAM TURBINE

The more modern method of extracting mechanical energy from heat energy is the steam turbine. Steam turbine have been the norm in various land based power plants for many years. BC Hydro's Burrard Thermal Plant just outside Vancouver is very similar to many power plants in most countries, and a good example of a steam power plant. The Burrard Generating Station is a 950 MW conventional natural gas-fired generating station. It's large boilers create large amounts of steam which is then fed to steam turbines. The turbines rotate large alternators, which produce electrical energy. On a ship, the operation is generally smaller, even on very large super tankers. On a ship, the turbine is connected to a reduction gear, which drives a propeller, producing motion instead of electrical energy.

If you can imagine a pinwheel, held solidly near your mouth, then blowing, at the right angle, air unto each "blade" of the pinwheel. You see the whole pinwheel turn. The principle of the impulse turbine is much the same.

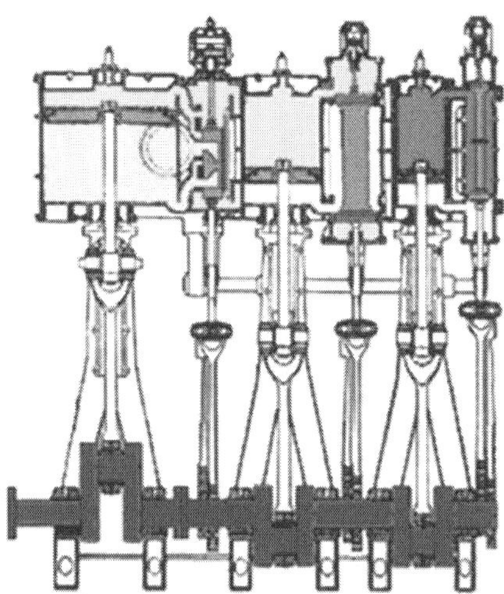

The impulse turbine contains several "pinwheels" which are actually called turbine rotors, pictured to the right. The rotors can rotate on a shaft, but cannot slide for and aft. "In front" of these rotors are nozzles, drilled into the stationary part of the turbine. Because steam does not like to be confined, each nozzle ejects steam onto one blade of the rotor, much like we imagined with the "pinwheel". Because the shape of the blades is at an angle, the jet of steam must change direction. This change in direction results in a force, rotating the rotor which rotates the shaft.

One set of turbine rotor and stationary nozzles is called a stage. Much like the triple expansion piston type engine, mentioned above, the steam travels through many stages. In the case of steam turbines, the steam proceeds through one stage, then collects and proceeds to the second stage and so on. Each time, the steam proceeds to a larger diameter rotor turbine, until the most of it's energy has been exerted on the rotors of the turbine. The energy depleted steam is drawn, by vacuum, to the condenserwhere it is cooled to form feed water, ready to feed the boiler once again.

As with any machine, improvements and specific designs have evolved to improve the overall efficiency of the machine. One turbine design is the impulse design as describe above. Another is the reaction type turbine, both types are illustrated below. A third is more of a hybrid design, combining, actually compounding, features from the impulse and reaction type steam turbines.

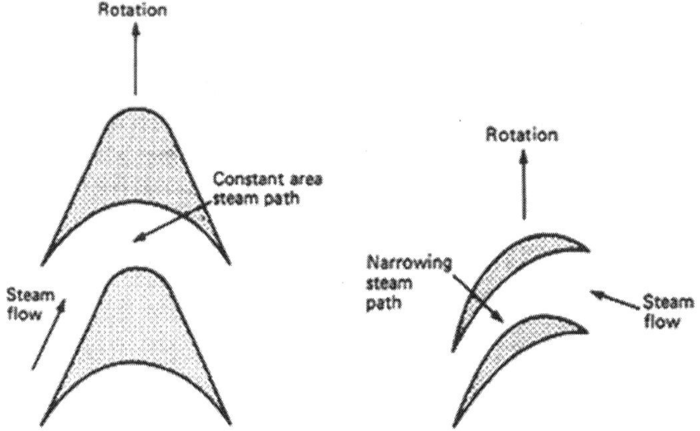

The impulse design (above left) relies on stationary ring of steam nozzles to direct flow onto the blades of a rotor. In the reaction type (above right), the flow of steam must pass through the rotor. The rotor is made up of blades, just like the impulse type, but in this case the blades are curved to provide a slight nozzle shape.

The blades on the impulse type change the direction of the steam, whilst in the reaction, the blades become the nozzles. The illustration above show the differences between the two types. The images to the right, courtesy of Rick Boggs' Merchant Marine and Maritime Pages, illustrates a reaction type steam turbine. Steam turbines rotate at very high speeds but in order to get the most efficiency from the propeller, the propeller must turn slow. Therefore a marine gear must be used. Marine gears are very common place, they are used to transform power from an engine to the actual machine doing the work, in this case the propeller.

In the picture to the left, the gear is situated behind the turbine. This is a 20,000hp steam turbine package. You can notice the condenser just below the turbine assembly.

THE TURBO-GENERATOR

The most obvious field for such a high-speed prime mover as the turbine was in the driving of electric generators, and it was for this purpose that it was originally designed. The dynamos of those days were small machines driven usually at 1000 to 1500 revolutions per minute by a belt from the flywheel of a reciprocating engine. Parsons required a dynamo that could be driven directly by his turbine at a speed of 18,000 revolutions per minute, in order that the combination should constitute a small, simple and self-contained generating unit. None of the established dynamo makers would have considered for a moment the construction of a machine so completely outside the range of previous experience.

It must be remembered that at the time electrical engineering was in a very elementary condition and dependent mainly upon empirical knowledge, for two years had still to elapse before Hopkinson propounded the theory of the magnetic circuit and laid down the fundamental principles of the design of electrical machinery. Parsons faced the question with the same boldness that he showed in the design of his turbine and he achieved an equally striking success. Both electrical and mechanical problems had to be solved, for the alternations of magnetism in the core were vastly more rapid than in any machine yet built, while the mechanical stresses to be provided against will be realised from the fact that a centrifugal force of 5.5 tons was developed by every pound of metal at the surface of the armature.

The dynamo was of the bi-polar type with an output of 75 amperes at 100 volts. Both turbine and dynamo fulfilled the anticipations of their designer, and after many years of useful work this historic unit was presented to the Science Museum at South Kensington, where it is carefully preserved for the instruction of future generations.

It required much persistence on the part of Parsons before his turbo-generators were able to enter their proper field of central station work. By 1888, although about two hundred of them were in service, they were employed almost exclusively for ship-lighting duties, and no electric light company had yet taken any notice of them. Parsons therefore decided that he would, himself, have to effect the introduction of the turbine into the industry that it was destined to dominate; so, aided by friends, he founded the Newcastle and District Electric Lighting Co., which began operations in January 1890 with a station at

Forth Banks equipped with a pair of 75 kw. turbo-alternators. But even this demonstration of its suitability for power-station service failed to arouse any general interest, and Parsons had to accept a financial risk in the businesses of companies formed to supply electricity to Cambridge in 1892 and to Scarborough in 1893 in order to give them sufficient confidence to install turbine machinery.

Progress thenceforward was rapid. The success of the turbine in saving the chief London station of the Metropolitan Electric Supply Co. from being shut down altogether in 1894 on account of the nuisance caused by its reciprocating engines attracted general attention and definitely established its footing in the industry. Larger and larger units were continually called for, and with every increase in size the advantages of the turbine became more apparent. Parsons' earlier machines had been constructed at the works of Messrs Clarke, Chapman and Co., in which firm he was a partner, but in 1889 he founded the present firm of Messrs C. A. Parsons and Co., Ltd., at Heaton, near Newcastle, in order to have complete control over their manufacture.

Parsons would have attained high fame for his electrical work alone had not this been- overshadowed in the minds of the public by the spectacular developments of his steam turbine. By 1900 he was building generating sets of 1000 kw. capacity, while in 1912 he undertook to build a turbo-alternator with an output of 25,000 kw., by far the largest and most efficient generating unit in the world at the time. This machine was installed in the Fisk Street Power Station of the City of Chicago, and it proved so successful that in 1923 Parsons was entrusted with the contract for a unit of 50,000 kw. for the same city. He lived to see an output of more than 200,000 kw. delivered by a single turbo-generator, and the reciprocating steam engine completely superseded by the turbine for central station work.

The growth of electricity supply consequent upon the invention of the turbine created a demand, not only for larger generating units, but also for higher transmission voltages in order that more extensive areas might be economically served. In the early days the practice had been to generate at about 2000 volts, and to step up this pressure when required by means of transformers. Ferranti had given a lead in the direction of higher generating voltages in 1889 by designing large slow-speed alternators to generate single-phase current at 10,000 volts for his famous Deptford Station. These machines were, however, recognised as exceptional and they had little or no influence on the industry generally.

The first real advance towards modern conditions was made by Parsons in 1905 when he supplied a pair of 1500 kw. turbo-alternators generating at 11,000 volts to the Frindsbury Power Station in Kent. Once it had been demonstrated that high-speed alternators could be safely constructed for this voltage, it soon became a usual generating pressure and remained so for many years. As before, when higher pressures were required for transmission they were obtained by the use of transformers, which were commonly attached permanently to the machines they served. There was, however, to Parsons' mind, something illogical in generating at 11,000 volts or thereabouts when the whole current might have to leave the station at a higher voltage. He therefore attacked the problem with his usual energy and insight, with the result that in 1928 he produced a 25,000 kw. Turbo-alternator designed to generate directly at 36,000 volts.

This was installed in the Brimsdown Power Station the same year. The windings of the machine were constructed in accordance with an entirely new principle, which made it possible to generate at 36,000 volts without submitting the insulation of the windings to any greater electrical stress than is usual in an ordinary 11,000 -volt generator. The machine was entirely successful and once more Parsons had set a new standard in power-station machinery. Many of the most important power stations, both in Great Britain and abroad, have now adopted the practice of generating directly at 36,000 volts, thereby eliminating the large and costly step-up transformers necessary with the previous method.

Enough has been said to indicate, in some small measure, how much the electrical industry owes to Parsons. He not only provided it with the turbo-generator, but led the way for more than a generation in every important development of power-station machinery. Excepting the introduction of the cylindrical form of rotor for turbo-alternators by the late C. E. L. Brown, who was building turbine machinery on the Continent under Parsons' patents, it may fairly be said that there was no notable improvement in the design of high-speed electrical machines that did not originate in the Heaton Works. Moreover, whenever a larger type of machine was called for, Parsons was ready to construct it, even if far beyond the capacity of anything previously built, provided only he was satisfied that the requirements could be successfully fulfilled. His enterprise was never restrained by considerations of what had been done, but only by what could be safely accomplished with the materials of the day. Typical of his engineering courage was the jump from 350 to 1000 kw. in 1900, and the still more spectacular leap from 6000 to 25,000 kw. in 1912. He was an equally great pioneer in all matters pertaining to efficiency.

As long ago as 1900 he made the first practical experiments with regard to the reheating of steam in the course of its expansion, and later proved the benefits of this procedure in the important power stations of North Tees, Barking, and Dunston, the last of which held for a time the record of efficiency for all British stations. It was Parsons also who introduced the now universal practice of extracting air from a condenser by means of a steam-jet, his 'Vacuum Augmentor' of 1902 being the direct prototype of the modern steam-jet air-ejector, without which the present efficiency of condensing plants would be impossible.

Parsons, again, was the first engineer to take practical advantage of the possibility of effecting an improvement of the thermodynamic cycle of a steam turbine plant by the regenerative heating of the feed water, a development for which he acknowledged his indebtedness to the original proposal of Mr James Weir in 1876. He applied this principle in the Blaydon Burn Power Station in 1916 by heating the feed water progressively by means of partially expanded steam extracted from the turbine at different pressures, a procedure which has since been adopted as an indispensable feature of every efficient steam power plant in the world. The result of these advances, coupled with innumerable other improvements due to his prolific brain, enabled him to construct generating units capable of operating with a heat consumption of no more than 9280 B.T.U.per kw. hour, a figure that even to-day could hardly be surpassed by the largest and most efficient machines in existence.

Although the technical merits of Parsons' work in the development of turbines and electrical generators can only be fully appreciated by experts, the benefits that have accrued from it are obvious to all. It is sufficient to contemplate the part played by electricity in our domestic and industrial well-being to realise how greatly we are dependent upon a cheap and abundant supply. This has come to be regarded as one of the necessities of civilised life, and it is very certain that the service we now enjoy would have been utterly impossible without the turbo-generator.

The cheapness of electricity produced by a steam power station depends mainly upon three factors, namely the quantity of fuel required to generate it, the capital charges on the station and equipment, and the expenses of running and maintaining the plant. On all of these the influence of turbine machinery has been profound. A modern station can be operated with a fraction of the fuel that would have been necessary for the same output in the days of the reciprocating engine, chiefly because the turbine can take advantage of a far greater range of expansion of the steam. The capital and labour charges are less because of the larger sizes of individual turbine units, while the maintenance costs are much reduced on account of the much greater simplicity and reliability of turbine machinery. The economy of fuel that has been due to the work of Parsons is incalculable. In the power stations of Great Britain alone the saving amounts to many millions of tons of coal per annum.

OTHER USES OF THE STEAM TURBINE ON LAND

Although the steam turbine finds its greatest field of usefulness on land in power stations and in driving the electrical generators in private plants, it has many other industrial applications. Turbines were employed at a very early date for driving centrifugal pumps, fans and blowers, all of which were naturally suitable for direct operation by a high-speed prime mover. Parsons also realised that if a turbine were driven by power, instead of being used to produce it, the machine could be used as a compressor. Many compressors for air and gas were constructed by him on the principle of the reversed axial-flow turbine. Better results, however, appeared at the time to be obtainable by machines working on the centrifugal principle, so that the axial-flow compressor fell into disuse. The researches in aerodynamics carried out in recent years have now led to a better understanding of the action of the blades in an axial compressor, with the result that the earlier difficulties have been overcome, and the present tendency is to revert to the type of machine originated by Parsons, especially when the highest efficiency is imperative.

The perfection of toothed gearing, to which Parsons contributed so greatly by his invention of the 'creep' system of cutting the teeth of gearwheels, opened up the whole industrial field to the turbine, as it was then no longer confined to the driving of such high-speed machinery as could be directly coupled to it. It was successfully applied even to the driving of steel rolling mills and other duties of a similarly exacting nature, and was often used to replace ordinary steam engines for driving the main shafts of factories. Even when reciprocating engines were retained, the ability of the turbine to work with steam at very low pressures was frequently taken advantage of by installing turbines to develop extra power from the exhaust steam of the engines which had hitherto been blown to waste.

Under other circumstances turbines could be used to generate the whole of the power required, and supply at the same time any desired amount of partially expanded steam at a given temperature and pressure for heating and process work. In all these developments Parsons took a leading part, providing turbine machinery to meet the most diverse requirements and thereby enabling factories and industrial undertakings to produce their own power much more cheaply and efficiently than before.

THE STEAM TURBINE AT SEA

The use of the steam turbine for the propulsion of ships was among the claims made by Sir Charles Parsons in his original patent of 1884, but he confined his energies at first to the task of establishing the position of the turbine on land, and it was not until 1894 that he took steps to apply it to marine duties. His works at Heaton were then so fully occupied with turbo-generators that he decided to establish a separate organisation, with works at Wallsend-on-Tyne, to deal with the special problems involved in marine propulsion. This company, which became known later as The Parsons Marine Steam Turbine Co., Ltd., proceeded immediately with the construction of a little vessel whose fame is now historic.

This was the *Turbinia*with a length of 100 feet and a displacement of 44 tons. After much experimental work with her propellers, the *Turbinia* attained a speed of 34 knots, which was a very remarkable achievement, since the fastest destroyers of the time could hardly exceed 27 knots. The fact that the steam turbine was inaugurating a new era in marine practice was brought home to the public in an unmistakable manner at the great Naval Review held in 1897 to celebrate the Diamond Jubilee of Queen Victoria. A vast fleet, representing not only the might of the British Navy, but the sea-power of other leading nations as well, was assembled off Spithead when the little *Turbinia*, with Parsons himself in control of the machinery, created a sensation by racing down the lines of warships at a speed obviously greater than that of any other vessel afloat. The Admiralty could not ignore such a demonstration, and entrusted Parsons with the construction of a 30-knot turbine-driven destroyer, H.M.S. *Viper*, but so grudgingly was the order given that Parsons and his associates were required to deposit a sum of no less than£100,000 as a security, in case the vessel should not come up to expectations.

These, however, were more than fulfilled, the *Viper* attaining a speed of over 37 knots when officially tested over the measured mile with turbines developing 12,000 H.P. A second turbine-driven destroyer, built about the same time, was taken over by the Admiralty and added to the Navy under the name of H.M.S. *Cobra*; but shortly afterwards both of these boats were lost at sea by accidents quite unconnected with the nature of their machinery, so that the little Turbinia became once again the only representative of the turbine principle afloat.

The lives of the *Viper* and *Cobra* had been brief, but their performance had attracted the attention of certain enterprising engineers connected with the Merchant Service, and in 1901 the first turbine-driven passenger vessel, the *King Edward*, was built for service on the river Clyde. This was followed by the *Queen Alexandra* for the same duties, and within the next year or two, turbine propulsion had also been adopted for the cross-channel boats *Queen* and *Brighton*. Meanwhile, in order that the advantages of turbines for warships should once more be demonstrated to the Naval authorities, The Parsons Marine Steam Turbine Co. laid down another turbine-driven destroyer which was eventually acquired for the Fleet under the name of H.M.S. *Velox*. The Admiralty now began to take the turbine more seriously, and when, in 1902, orders were placed for four 3000-ton cruisers, it was decided that one of them, H.M.S*Amethyst*, should be fitted with turbines in order that a comparison might be made between her performance and that of the three sister vessels equipped with the usual reciprocating engines. The results were so conclusively in favour of the *Amethyst* that the last prejudices against turbine machinery in the Navy were overcome and the way was open for its general adoption.

The first turbine vessel to cross the Atlantic was the steam yacht *Emerald*, built in 1909 to the order of Sir Christopher Furness, but the next year the Allan Line ordered two 13000-ton vessels, the *Virginian* and *Victorian*, for their Liverpool-Canada passenger service, and with praiseworthy enterprise they decided that they should be propelled by turbines. The Cunard Company followed with a 30,000-ton liner, the *Carmania*, which once more demonstrated the superiority of the turbine by proving, on her trials in 1905, fully a knot faster than her sister ship the *Caronia*, equipped with reciprocating engines. By this time Parsons had won his battle for the recognition of the turbine at sea. His victory was confirmed, so far as Naval vessels were concerned, by a Committee on Naval Design appointed by the Admiralty in 1905, who advised that in future turbine machinery should be used exclusively in all classes of warships. As a consequence of this decision, turbines were adopted for the propulsion of H.M.S. *Dreadnought*, the fastest and most powerfully armed battleship in the world at the time.

The greatness of the contribution that Parsons had made to Naval progress will be understood from the words used by the First Lord of the Admiralty in justification of the decision to adopt turbine machinery. The turbine system, he said, had been decided on 'because of the saving in weight and reduction in the number of working parts and reduced liability to breakdown; its smooth working, ease of manipulation, saving in coal consumption at high powers, and hence in boiler-room space, and saving in engine-room complement; also because of the increased protection provided with this system, due to the engines being lower in the ship'.

Most of the reasons which caused the Admiralty to renounce reciprocating engines in favour of turbines in Naval vessels applied with equal force to a large part of the Mercantile Marine. Indeed, the conditions of service of fast liners, required to make uninterrupted passages at full speed across the ocean, enabled the turbine to make an even more advantageous showing than in war vessels, which were rarely required to navigate at full speed. In 1904 the British Government came to an arrangement with the Cunard Company, under which the latter should construct two new liners with an average speed of at least 24.5 knots, in order to be serviceable, not only as fast mail carriers but also as Naval auxiliaries in the event of war. The question of their propelling machinery was referred to a strong Commission representative both of the Admiralty and the leading shipbuilding firms, who reported definitely in favour of the turbine. This decision, coupled with that of the Admiralty to adopt turbine propulsion exclusively in the Navy, shows how enormous had been the change in practice since the first appearance of the *Turbinia* only ten years before.

The two new liners, the *Lusitania* and *Mauretania*, were launched in 1906 and went into service the following year. The vessels were practically identical in design. The *Mauretania* had a displacement of 38,000 tons, and obtained a speed of 26.04 knots on her 48 hours' full-power trials with her turbines developing 70,000 horse-power. She captured the 'Blue Ribbon of the Atlantic' for the fastest crossing and held this honour for nearly a quarter of a century. Her sister vessel, the *Lusitania*, will be remembered as having been sunk without warning by a German submarine in 1917 with the loss of over a thousand passengers and crew. The *Mauretania* was taken out of commission in 1935 and broken up after a working life of 28 years. Her 70,000 H.P. by no means marked the limit of power of marine installations. In the Royal Navy H.M.S. *Hood* was constructed during the last war with turbines of 150,000 H.P., and even this power exceeded in recent Atlantic liners such as the *Queen Mary* and *Queen Elizabeth*.

Enough has been said to show that the success of the turbine at sea was no less striking than its achievements on land, and its supremacy over the reciprocating engine for all the most important classes of service was even more rapidly established. It permitted vessels to be driven at speeds that had previously been impossible, and enabled those speeds to be maintained in the roughest of seas. In warships the turbine machinery could be protected more effectively than engines, the fuel economy was greater and the maintenance costs less. Commercial vessels benefited similarly in speed and economy by turbine propulsion, while the space available for cargo and passengers was increased and vibration was lessened. The cumulative effect of these advantages was sufficient to establish the turbine, within a few years only, as the recognised prime mover for all the Navies of the world, as well as for all the fastest ocean liners. The subsequent introduction by Parsons of gearing between the turbines and the propellers was another great step in advance, for it not only diminished the size of the machinery and increased its efficiency, but it enabled the ordinary cargo vessel to profit equally by the employment of turbines.

The turbine, however, has never altogether succeeded in ousting the well-tried marine engine from slow-speed cargo vessels. On the contrary it did much to give the engine a new lease of life, for by the addition of a turbine to develop power from the exhaust steam of the engines, the efficiency of the machinery was considerably increased. A combination of this kind was patented by Parsons in 1906, and was first used on a commercial scale in the 10,000-ton S.S. *Otaki* in 1908. This vessel showed a reduction in service of 12 per cent in fuel consumption as compared with her sister ships equipped with engines only, the saving amounting to 750 tons of coal on the round voyage to New Zealand and back.

The *Otaki* was a three-shaft vessel with the turbine alone driving the centre shaft, but in subsequent developments it has been the general practice to couple the turbine through gearing to one of the engine-driven shafts. This arrangement, in one form or another, has been exploited by various Continental manufacturers, who have associated their names with particular variants of it, but the credit of it belongs properly to Parsons. It was stated on good authority in 1926 that the efficiency of machinery for the propulsion of ships had been more than doubled during the twenty years that had then elapsed since the introduction of the turbine, the improvement being due in no small measure to the use of gearing. Progress has, of course, continued, although it may not have proceeded subsequently with the same rapidity, and the consequent aggregate reduction in the quantity of fuel consumed by the Mercantile and Naval fleets of all nations due to the work of Parsons has been incalculable.

The insistent desire of shipowners for an even greater economy of fuel led to the development of the marine oil engine, which, after the end of the war of 1914-18, began to challenge the supremacy of turbine machinery, particularly for mercantile vessels of slow and moderate speed. The turbine, however, had the advantage of much greater mechanical simplicity, and Parsons sought to bring its fuel consumption more nearly into line with that of the oil engine by urging the adoption of higher steam pressures and temperatures at sea. Although marine engineers had been converted to a belief in turbines and gearing, they had always been conservative in the matter of boiler practice. In 1926 conditions in the mercantile marine did not much exceed a pressure of 200 lb. per sq. inch and a temperature of 500° F. In Naval work matters were somewhat better, but not much, as the steam pressure in warships was only about 275 lb. Practice on land was very much more advanced.

At that time many central stations were already using steam at 500 or 600 lb. pressure, superheated to 750° F. or over, while in some cases pressures of the order of 1400 lb. per sq. inch had been adopted. Parsons felt very strongly that marine engineers ought to take advantage of the economies resulting from higher pressures and temperatures. Knowing that a practical demonstration was the surest and quickest way of convincing the sceptics, he arranged for the equipment of a small passenger vessel, the *King George V*, with geared turbines of 3500 H.P. to work with steam at 550 lb. per sq. inch, superheated to 750° F. The steam was supplied by water-tube boilers of the Yarrow type. The *King George V* was the pioneer of high-pressure steam at sea. Although the installation was a comparatively small one, and the conditions of a river steamer making short trips with frequent stops were not the most favourable for the experiment, the machinery fulfilled the expectations of its designers, the full-load trials made after a short period of commercial service showing a steam consumption of only 8.01 lb. per shaft horse-power hour of the turbines. This enterprise of Parsons once more opened up a new field for marine engineers, and higher pressures at sea soon became general. Within the next few years there were a number of Atlantic and Pacific liners operating with steam at 350 and 400 lb. pressure, and in 1931 the Admiralty gave their approval to the advance in steam conditions by adopting a boiler pressure of 500 lb. per sq. inch and a temperature of 750° F. in H.M.S.*Acheron*. This vessel had a consumption of only 7.7 lb. of steam and 0.608 lb. of fuel per shaft horse-power hour, which constituted a record for economy in Naval work. Mercantile practice has now attained rivalry with Naval practice by the use of steam at an even higher pressure in the latest Cunard liners.

PARSONS' WORK ON SCREW PROPELLERS

The application of the steam turbine to marine propulsion gave rise to many incidental problems, by the solution of which Parsons made notable contributions to the progress of marine engineering. One of the earliest difficulties he encountered was due to the high speed of the propellers. The first machinery of the *Turbinia* consisted of one turbine driving a single propeller at 2000 R.P.M. The results of the trials were disappointing. Different designs of propeller were tried but the best speed that could be obtained was only about 20 knots. It was clear, either that the turbine was not developing its rated power, or that the efficiency of the propeller was extremely low. To settle this question Parsons devised a special apparatus to measure the torque exerted by the turbine on the propeller shaft. This instrument was the prototype of the modern torsion meter, and by its use he assured himself that the fault was in the propeller and not in the turbine. About the same time similar difficulties in obtaining the anticipated speed were experienced in a new class of very fast torpedo-boats which were fitted with reciprocating engines. Both Parsons and the Naval authorities arrived at the same conclusion, namely, that the trouble was caused by the inability of the water to follow the rapidly moving propeller blades, so that a vacuous space was left behind the blade tips, with a consequent loss of propulsive power. This phenomenon, now known as 'cavitation', is also liable to occur in centrifugal pumps and water turbines when conditions are favourable to it. Parsons met his immediate difficulties by providing the *Turbinia* with three shafts each carrying three propellers so that the whole propulsive power was divided among nine propellers. With this alteration the vessel attained a speed of over 34 knots.

Many men would have rested content to have successfully circumvented their difficulty, but Parsons realised the importance of a thorough investigation of the whole question of cavitation, as this was clearly going to be a matter of concern to designers of high-speed vessels. He therefore constructed a tank with glass sides in which a model propeller could be run at high speeds. The propeller was strongly illuminated by intermittent light, the speed of the flashes being regulated in accordance with the revolutions of the propeller so that the blades could be made to appear stationary or only revolving very slowly. It was recognised that cavitation would be favoured by working with water near its boiling-point, so the first experiments were made with hot water. It was found, however, more convenient to attain the same result by maintaining a vacuum above the water in the tank, and the nature of cavitation was exhaustively studied in this manner. The knowledge gained by these investigations led to great improvements in the design of high-speed propellers, and the methods of study initiated by Parsons have since become generally adopted.

Closely allied with the phenomenon of cavitation is that of the erosion of propeller blades, although the connection between the two was not at first realised. Erosion had become such a serious problem that in 1915 the Admiralty appointed a Committee to report on the subject. In view of Parsons' experience of propeller design, he was requested to serve on the Committee, and it was he who suggested that the erosion was probably a secondary effect of cavitation. His view, which is now generally accepted, was that the vacuous spaces typical of cavitation were continually collapsing, causing a hammering by the water on the metal of the propeller. This. hammering might easily attain a destructive intensity owing to the absence of any appreciable quantity of air or gas in the cavity to soften the blows.

It was typical of Parsons that he would accept no theory, not even own, that could not be supported by experiment, so he set himself to test his idea. The method he adopted was as simple and direct as it was ingenious. He made a hollow brass cone with a small hole in its apex which could be closed by a plate of the metal to be operated on. This cone was held face downwards in a tank, and when filled with water it was forced suddenly downwards until arrested by a rubber cushion on the bottom of the tank. The resilience of the rubber permitted the water in the cone to continue its downward motion for a moment after the cone had stopped, thus causing a vacuous space to occur at the top of the cone. This space immediately collapsed, and the returning water was found to strike the plate with a force often sufficient to puncture it. Pressures as great as 140 tons per sq. inch were obtained in this way, and the results of the experiments left no doubt that the damage met with in propeller blades could be fully accounted for by the hammering action consequent upon cavitation, as suggested by Parsons.

MECHANICAL GEARING FOR MARINE AND LAND TURBINES

The steam turbine is essentially a high-speed prime mover, and it therefore shows to its best advantage when directly coupled to machinery that can be run at the economical speed of the turbine. This speed is, fortunately, suitable for a large range of electrical machines, but the smaller sizes of alternators and most continuous current generators require to run at less than the optimum turbine speed, which indeed may be altogether too high to make direct driving advisable or even practicable in many instances.

The obvious way of arranging for the speed of the turbine to be independent of that of the driven machinery is by interposing speed-reducing gear between' the two. The use of gearing in connection with steam turbines was originated by Dr de Laval in Sweden about the year 1889. He used it with great success in the small single-wheel turbines associated with his name, but Parsons was able to develop his turbines without recourse to gearing because of their lower efficient speed. He did, however, construct a geared unit in 1896, consisting of a 150 kw. alternator running at 4800 R.P.M. driven by a turbine running at 9600 R.P.M., and the next year he equipped a small steam launch with a turbine running at no less than 19,600 R.P.M. and driving two propeller shafts at 1400 R.P.M. by means of single-helical gear-wheels. Both these installations were quite successful, but for the next ten years or so Parsons made no further applications of gearing to either land or marine turbines. By that time he had firmly established the position of the steam turbine at sea. The *Dreadnought, Lusitania* and *Mauretania* were already in service, and the performances of these and other vessels had demonstrated the great superiority of the turbine over the reciprocating engine for the propulsion of all warships and the fastest mercantile vessels. But there remained a large field still to be conquered. The immense fleets of slow-speed tramp steamers and cargo ships on all the seven seas were unable to benefit from turbine propulsion because the normal speeds of their propellers were too far below the rotational speeds desirable for turbines.

The reason for the discrepancy between the most efficient speeds of a turbine and a propeller lies in the enormous difference in the density of the media-steam and water-in which they are respectively working. A turbine can only be made to run slowly with efficiency, either by constructing it with a very large diameter in order to maintain the peripheral speed of the blades, or by using a very large number of blade rows so as to reduce the steam velocity per stage. In either case the dimensions become excessive. Parsons realised that the only real solution to the problem was to be found in providing some connection between the turbine and the propeller shaft which would enable each to run at the speed most conducive to efficiency, and he therefore turned his attention again to the possibilities of mechanical gearing. He commenced by carrying out exhaustive experiments to determine what tooth-speeds could be employed and what power could be transmitted consistently with safety and durability. The results were so encouraging that new possibilities were opened up for turbine machinery both on land and sea. To make a practical test of the use of gearing in marine work, The Parsons Marine Steam Turbine Co. purchased in 1909 an old cargo vessel, the *Vespasian*, of 4350 tons displacement, and replaced her 750 H.P. triple expansion engines by geared turbines. The success of the experiment has become historical. With the same boilers and steam pressure, the substitution of geared turbines for the original machinery resulted in a reduction of the fuel consumption by 15 per cent. The gearing worked perfectly, and after the *Vespasian* had completed several years of commercial service, her hull, which was then worn out, was broken up and the turbines and gearing transferred to another vessel.

By 1919, or only ten years after the first experiments with gearing in the *Vespasian*, it was estimated that no,000,000 H.P. were being transmitted through gearing in warships and merchant vessels, and as much as 25,000 H.P. had been transmitted by a single gear-wheel. The introduction of gearing in connection with marine turbines gave rise to a new problem. So long as it was the practice to couple turbines directly to the propeller shafts, it was possible to arrange that the thrust of the propeller should be largely counterbalanced by the axial pressure of the steam on the rotor blading, so that only a small differential pressure had to be carried by the thrust-block. With geared turbines no such counterbalancing was possible, and the thrust-block had the duty of transmitting the whole of the propeller thrust to the structure of the vessel. The old multi-collar thrust-block had served well enough with reciprocating engines, and had indeed been retained in the *Vespasian*, but the type was no longer adequate for the higher shaft speeds of turbine-driven vessels generally. Fortunately, about this time, the pivoted-pad type of thrust-block, in which the whole of the end-thrust could be carried on a single collar, was invented by Mr A. G. Michell in Australia. It was an engineering novelty and had only been made in quite small sizes, but Parsons at once realised its potentialities for marine work. He therefore constructed an experimental thrust-block on the Michell principle, designed to work with an axial pressure of 40,000 lb. and large enough for the propeller shaft of a destroyer. The result was so satisfactory that this type of thrust-block became a standard,component of all geared marine turbine installations.

During the war of 1914-18 single-collar pivoted-pad thrust-blocks were fitted to Naval vessels totalling 10,000,000 H.P., and the construction of such a ship as H.M.S. *Hood*, in which 36,000 H.P. had to be transmitted through each of the four propeller shafts, would have been impossible without this type of thrust-block.

Simultaneously with the application of gearing to marine purposes, an equally bold departure was made by Parsons in land practice. He supplied a 750 B.H.P. turbine running at 2000 R.P.M. for the onerous duty of driving a rolling-mill for the production of ships' plates. The rolls had to run at 70 R.P.M., and this speed was obtained by the interposition of double reduction gearing between the turbine and the mill. The plant proved a most gratifying success, and there could no longer be any doubt that the use of mechanical gearing would enable the turbine to drive the reciprocating engine from almost its last strongholds.

In order that gear-wheels should work quietly and without deterioration under the conditions of speed and power imposed by their new duties, it was of course essential that their teeth should be extremely accurate both as to form and pitch. In the ordinary method of gear cutting, every error of pitch that may exist in the master wheel of the gear-cutting machine will necessarily be reproduced in the wheel being cut. No master wheel can be mathematically perfect, and the accuracy desired by Parsons was greater than any ordinary gear-cutting machine could provide. He therefore turned his attention to the production of better gears, and to this end he devised what is known as the 'Parsons Creep Mechanism'. By this mechanism the work-table of the machine was caused to rotate slightly faster than the master wheel, with the result that any errors existing in the latter were distributed spirally round the wheel being cut, instead of being concentrated at one part of the circumference. The consequence was that the unavoidable defects of the master wheel were, for all practical purposes, completely eliminated in the work.

This method of 'creep-cutting', invented by Parsons in 1912, created an entirely new standard of accuracy for mechanical gearing, and made it possible to produce gear-wheels that could be relied on to transmit any desired power with quietness and durability. Thenceforward the turbine was free from all limitations imposed by the speed of the driven machinery, for each could be run at its most efficient rate, the connection between the two being made by appropriate gearing. Geared turbines were soon employed as the propelling machinery for steam ships ranging from slow cargo vessels to the fastest warships and liners, for even in the case of fast ships it was recognised that high-speed turbines and gearing were more economical in service than direct-coupled machines running at a speed dictated by the requirements of the propeller.

The improvement brought about by gearing may be illustrated by the following comparison. With the early direct-coupled marine turbines the steam consumption was about 15 or 16 lb. per shaft horse-power hour, while by 1923 geared installations could operate with a consumption of less than 10 lb. of steam for the same power, which could be reduced to 8 lb. or less if the steam was superheated. If we take into account the simultaneous increase of the efficiency of the propeller due to its lower speed, it is fair to say that by the introduction of gearing Parsons effected a further economy in fuel comparable with that originally brought about by the application of the turbine to marine work.

On land there was not the same scope for geared turbines as in marine practice, but the introduction of gearing has nevertheless widened the field and improved the performance of turbine machinery in many directions. Continuous current dynamos, centrifugal pumps and small alternators have in general to be run at speeds considerably below the economical speeds of small turbines. By the incorporation of gearing into the unit, these and other machines can be driven by efficient high-speed turbines, thus leaving a very small field for the reciprocating engine in industry.

PARSONS' WORK ON SEARCHLIGHT REFLECTORS

The importance of the part played by Parsons in the development of efficient searchlights is better appreciated by Naval and Military technicians than by the general public. Searchlights were, of course, already in use at the commencement of Parsons' engineering career. They were employed by the British Navy in 1876, and their value was demonstrated in the Egyptian campaign of 1882. The advantage of using an accurately parabolic reflector to project the beam of light from the arc was recognized from the first, but owing to the difficulty of making a true parabolic shape, the earliest reflectors for searchlights were formed with spherical surfaces. The light was reflected from a coating of silver deposited on the back of the glass from which the mirror was made, and the glass was given an increasing thickness from the centre to the rim so that the consequent difference in refraction experienced by the rays should cause them to be projected in a fairly parallel beam. Large mirrors of this design were very heavy and expensive, and they were liable to become fractured in service owing to the differential expansion when subjected to the heat of the arc.

Parsons set himself the problem of producing at reasonable cost, a silvered reflector that should be of uniformly thin glass and of the ideal parabolic curvature. He worked out a process of manufacturing such reflectors while he was a partner in the firm of Clarke, Chapman and Co., of Gateshead, and as soon as he established his own Works at Heaton in 1889 he organised a special department for their production. By so doing' he virtually founded a new industry, and one that has contributed greatly not only to the needs of national defence, but also to those of peaceful commerce.

The first parabolic reflectors for searchlights did not exceed 30 inches in diameter, but to meet the increasing demands of the Naval and Military authorities, parabolic reflectors up to more than 7 feet in diameter were successfully produced at Heaton. Concurrently with these developments in size, Parsons introduced many improvements in construction. The silvered side of the glass was protected by a coat of copper deposited electrically, and later on further protection was given by an additional backing of sheet lead reinforced by wire netting. This rendered the reflector immune from damage by exposure to oil fumes, salt water and other destructive influences met with in service. It also greatly reduced the risks of breakage. Indeed, a mirror protected in this way may remain serviceable after being pierced by a rifle bullet or a shell splinter.

Parsons, however, did not confine his efforts to the perfection of parabolic reflectors for throwing straight parallel shafts of light. For certain purposes, as for example when a large area such as a harbour or the landing ground of an aerodrome has to be illuminated, what is required is a flat divergent beam. To produce such beams, Parsons invented and devised methodsfor the manufacture of a most ingenious form of reflector, curved to a parabolic form in the vertical plane and to an elliptical form in the horizontal plane, both curves having a common focus.

The parabolic curvature resulted in the light issuing in a beam of uniform depth, while the effect of the transverse elliptical curvature was to cause the rays first to converge into a vertical line at the secondary focus of the ellipse, and then to diverge at a predetermined angle. Searchlights equipped with such reflectors therefore not only projected a fan-shaped beam of the required type, but the whole of the light was able to pass through a narrow vertical slot situated at the secondary focus. Consequently the searchlight could be operated behind a loop-hole where its chances of being damaged by rifle fire would be very slight. Another of Parsons' inventions in connection with searchlights that has proved of the greatest value in navigation, particularly to ships passing through the Suez Canal, is the split parabolic-elliptical reflector. This is divided vertically into halves which are hinged to one another. Normally the combination acts as a single reflector, throwing one slightly diverging beam of light straight ahead from the ship. Such a beam enables the Canal to be safely traversed at night, but it would have a blinding effect on the pilot of any other vessel that might be approaching. It is to avoid this danger that the split reflector was devised. As soon as another ship is seen coming, the hinged halves of the mirror are swung apart by the manipulation of a small lever at the back, with the consequence that the beam is divided into two parts with a dark unlit space between them. The sides of the Canal are thus brilliantly illuminated, while the approaching pilot is not in any way inconvenienced by the light. Parsons constructed complete searchlight units of this kind, and in the early days they were hired out to ships at the entrance of the Canal and returned to shore at the other end for the use of ships making the reverse passage. This service enabled shipping to navigate the Canal during the hours of darkness, with a consequent saving of time and harbour dues.

STEAM TURBINES GEARING ARRANGEMENT

TURBINE GEARING

Steam turbines operate at speeds up to 6000rev/min. Medium-speed diesel engines operate up to about 750rev/min. The best propeller speed for efficient operation is in the region of 80 to 100 rev/min. The turbine or engine shaft speed is reduced to that of the propeller by the use of a system of gearing. Helical gears have been used for many years and remain a part of most systems of gearing. Epicyclic gears with their compact, lightweight, construction are being increasingly used in marine transmissions.

EPICYCLIC GEARING

This is a system of gears where one or more wheels travel around the outside or inside of another wheel whose axis is fixed. The different arrangements are known as planetary gear, solar gear and star gear

The wheel on the principal axis is called the sun wheel. The wheel whose centre revolves around the principal axis is the planet wheel. An internal-teeth gear which meshes with the planet wheel is called the annulus.

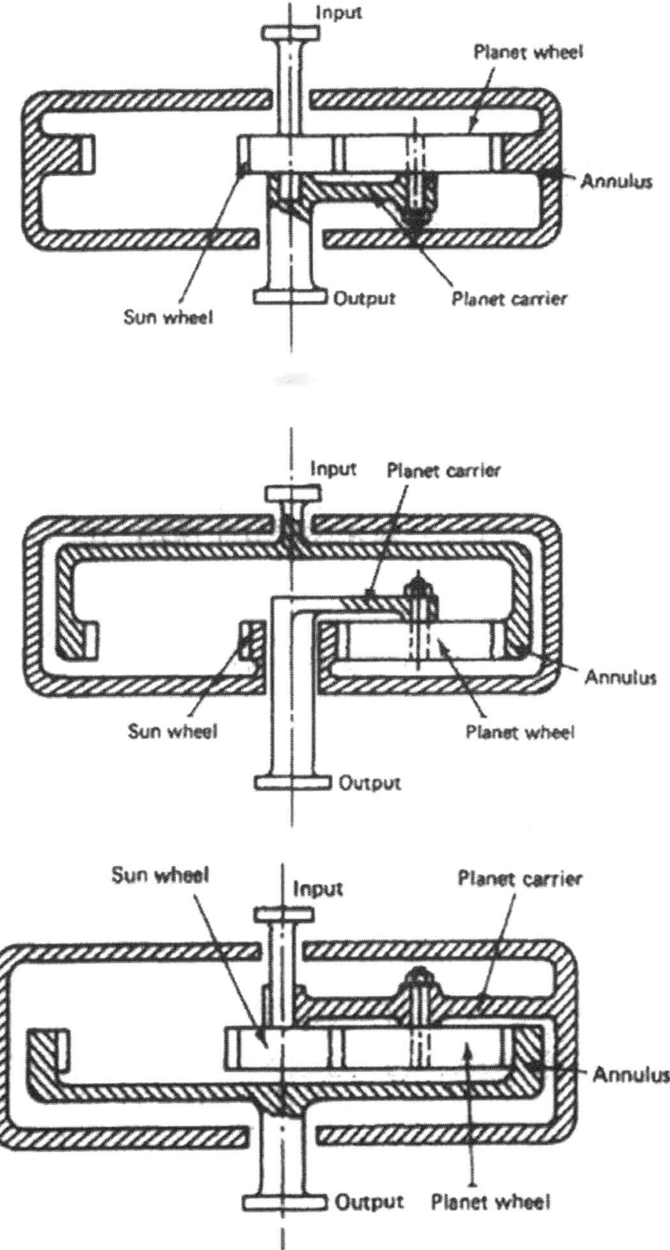

Fig. Steam turbine epicyclic gearing

The different arrangements of fixed arms and sizing of the sun and planet wheels provide a variety of different reduction ratios. Steam turbine gearing may be double or triple reduction and will be a combination from input to output of star and planetary modes in conjunction with helical gearing.

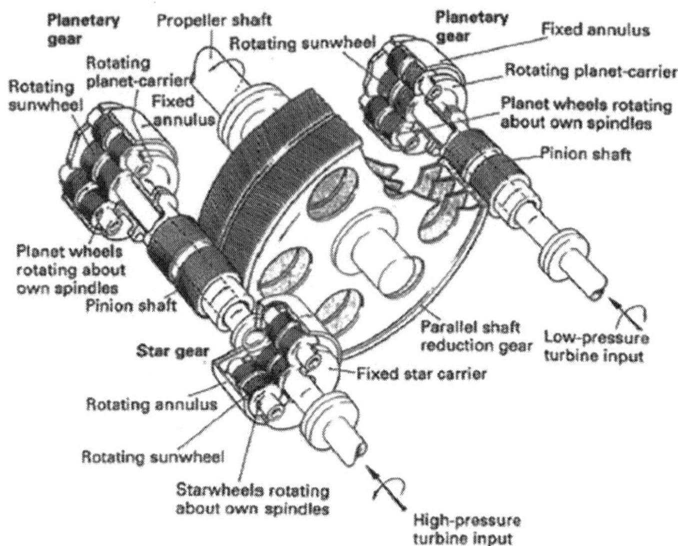

Fig. Turbine reduction gear

HELICAL GEARING

Single or double reduction systems may be used, although double reduction is more usual. With single reduction the turbine drives a pinion with a small number of teeth and this pinion drives the main wheel which is directly coupled to the propeller shaft. With double reduction the turbine drives a primary pinion which drives a primary wheel. The primary wheel drives, on the same shaft, a secondary pinion which drives the main wheel. The main wheel is directly coupled to the propeller shaft. A double reduction gearing system is shown in Figure below.

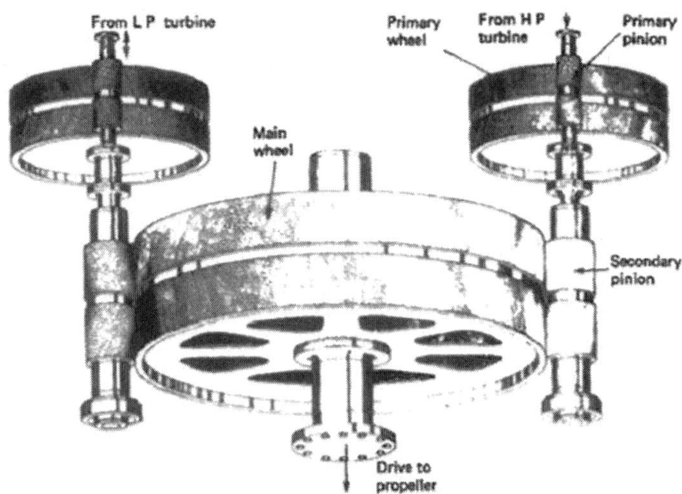

Fig. Turbine double reduction system

All modern marine gearing is of the double helical type. Helical means that the teeth form part of a helix on the periphery of the pinion or gear wheel. This means that at any time several teeth are in contact and thus the spread and transfer of load is much smoother. Double helical refers to the use of two wheels or pinions on each shaft with the teeth cut in opposite directions. This is because a single set of meshing helical teeth would produce a sideways force, moving the gears out of alignment. The double set in effect balances out this sideways force. The gearing system shown in Figure is double helical.

Lubrication of the meshing teeth is from the turbine lubricating oil supply. Sprayers are used to project oil at the meshing points both above and below and are arranged along the length of the gear wheel.

FLEXIBLE COUPLING

A flexible coupling is always fitted between the turbine rotor and the gearbox pinion. It permits slight rotor and pinion misalignment as well as allowing for axial movement of the rotor due to expansion. Various designs of flexible coupling are in use using teeth, flexible discs, membranes, etc.

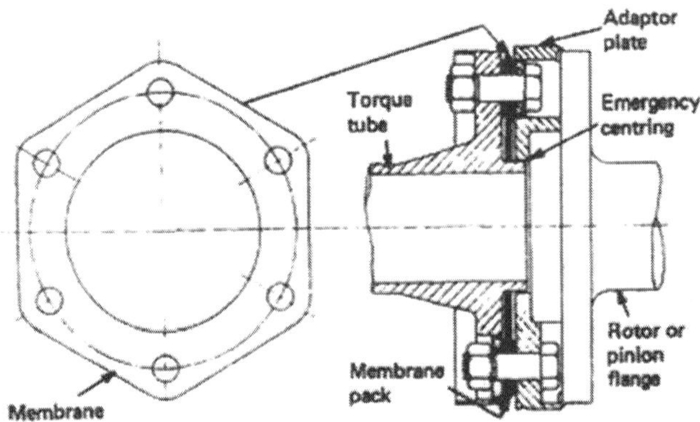

Fig. Turbine flexible coupling

The membrane-type flexible coupling shown in Figure above is made up of a torque tube, membranes and adaptor plates. The torque tube fits between the turbine rotor and the gearbox pinion. The adaptor plates are spigoted and dowelled onto the turbine and pinion flanges and the membrane plates are bolted between the torque tube and the adaptor plates. The flexing of the membrane plates enables axial and transverse movement to take place. The torque tube enters the adaptor plate with a clearance which will provide an emergency centring should the membranes fail. The bolts in their clearance holes would provide the continuing drive until the shaft could be stopped.

TURNING GEAR

The turning gear on a turbine installation is a reversible electric motor driving a gearwheel which meshes into the high-pressure turbine primary pinion. It is used for gearwheel and turbine rotation during maintenance or when warming-through prior to manoeuvring.

SAFETY DEVICES FITTED IN STEERING GEARS ON SHIPS

We have already studied about the general overview of steering gears and construction and components of a steering gear arrangement. We also saw how steering gear testing is carried out on board various types of ships. In this chapter we will continue to study about steering gears and talk about the various safety devices which are fitted on them such as the liquid level switch and other arrangements so as to ensure safe operation of the entire system.

SWITCH LEVEL 1:

The steering system has two lube oil tank systems with 2 rams functioning on each system. The level switch 1 gives an initial alarm following a loss of oil from either system. In normal operation one power unit provides hydraulic power to all four rams.

The causes for the loss of oil is mainly due to the broken hydraulic pipes and damage in the ram thus the leakage oil drains to the bilge well.

SWITCH LEVEL 2

When no action is taken immediately upon the previous alarm, the loss of oil continues and over a period of time this loss of oil initiates one or both of the level switches 2. This leads to the energization of a solenoid operated servo valve causing a combined isolating valve and bypass valve to operate. Thus splitting the system such that each power unit supplies power to 2 rams only.

At the same time this switch automatically starts the other standby helesaw pump to assist the turning and building up the pressure in the rams.

SWITCH LEVEL 3

This switch will get activated when there is still no further improvement in the loss of oil on the faulty side. In such a situation the steering continues un-interrupted if one unit is stopped and thus the rudder is turned with the help of the other unit system of 2 rams only, isolating the faulty one and this is called as emergency steering.

Thus the steering system is designed to work at half of the maximum torque on the steering system.

The defective system is put out of action and isolated.

RELIEF VALVE

Relief valve is fitted in the system to prevent over pressure in the hydraulic system due to shock loading of rudder. Shock loading of the rudder could occur in a variety of cases such as bad weather.

MANUAL BY PASS VALVE

This valve is only operated when there is a failure of the one of the systems and the rudder stock has to be turned mechanically, so that the ideal fluid can flow from the high pressure side to low pressure side.

LOW PRESSURE VALVE

Low pressure valve is a NON RETURN VALVE fitted in the main hydraulic pressure line, which opens for oil filling from a header tank if a low pressure situation occurs in the system.

ELECTRICAL STOPPER

It stops the hydraulic pump when the position of the rudder turning is at an angle of 36^0. It automatically cuts of the supply because it is the maximum turning angle of rudder. (It normally operates at 35^0 angle).

MECHANICAL STOPPER

In case of electrical stopper fails, the mechanical stopper is provided so that it stops the tiller arm after 36^0 angle of rudder

Electrical motor overload alarm

Whenever hydraulic motor or helesaw pump motor draws excessive current during bad weather, the steering should not fail. Instead it gives and alarm about excessive current being drawn by the motor. Continued recurrence of this alarm could mean that the electrical motor might burn

POWER FAILURE ALARM

The full power failure alarm will be raised if the power fails in black out condition. Emergency power should be arranged within 45 seconds and the emergency generator must be capable to give power for at least 30 minutes for big ship and 10 min in case of small ships (ships<10,000 ton displacement).

GEARS AND CLUTCH MECHANISMS

PRINCIPLES OF MARINE GEAR BOXES

Gearing is used in *drive trains* (sets of intermeshed gear wheels) to alter direction, position or speed/mechanical advantage of propulsion and auxiliary equipment such as winches, pumps and steering. The gear and shaft driven directly by the motor can be called the input and the final gear and shaft it drives can be called the output. Drive train A shown below illustrates reduction and reversal of the output, drive train B illustrates reduction only with further displacement (to the right) of the output, drive train C illustrates reduction and splitting into two outputs.

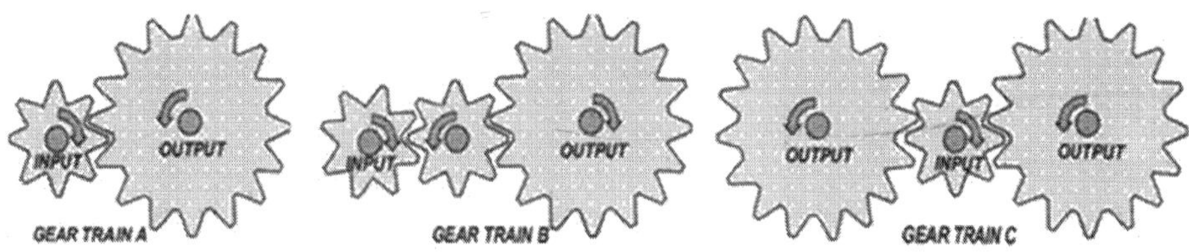

The efficient crankshaft speed of a small high speed motor is greater than the efficient speed of a propeller that revolves in relatively dense salt water. The solution to this is to use of a drive train with an input shaft and drive gear wheel enmeshed with a set of follower gear wheels connected to an output shaft (a gear box).

The output shaft follower (to the propeller) has more gear teeth than the input shaft driver (from the crankshaft), and therefore it rotates more slowly. The number of teeth in each gear wheel has a direct relationship to not only the speed of the follower but also the transfer of torque. The reduction ratios shown in the drawings above are calculated by the:

$$\frac{number\ of\ gear\ teeth\ (cogs)\ in\ the\ large\ wheel}{number\ of\ gear\ teeth\ in\ the\ small\ wheel} = \frac{16}{8} = 2\ or\ a\ \text{Re}duction\ Ratio\ of\ 2:1$$

Therefore if the engine speed is 1000 rpm and the propeller speed is 500 rpm then:

$$\frac{rpm\ of\ the\ engine}{the\ rpm\ of\ the\ propeller} = \frac{1000}{500} = 2\ or\ a\ \text{Re}duction\ Ratio\ of\ 2:1$$

While the speed has been halved, the mechanical advantage has been doubled (not allowing for efficiency losses due to friction and heat). Marine gear boxes are designed to reduce the speed of the propeller shaft but some applications require a greater speed from the output shaft.

This is simply achieved by using a smaller follower to give higher output speed with less mechanical advantage. It must also be noted that by adding a following gear wheel it will turn in the opposite direction to its driving wheel (in reverse).

GEAR SELECTION AND ENGAGEMENT

Gear box arrangements to select reverse gear include mechanically (using a gear lever directly or through a cable or linkage), electro-mechanically (using electrical solenoids which change the gears directly or hydraulically through clutches) or by hydraulic operation alone. The gearbox control and engine throttle (speed) may use separate levers for thrust and speed, or be combined in a single control lever as drawn below.

Marine transmission systems are heavy and generate considerable momentum so changing gear is smoothed by clutches. To avoid clashing of gears and overload on clutches the sequence of changing requires a delay for the propeller and shaft to slow down.

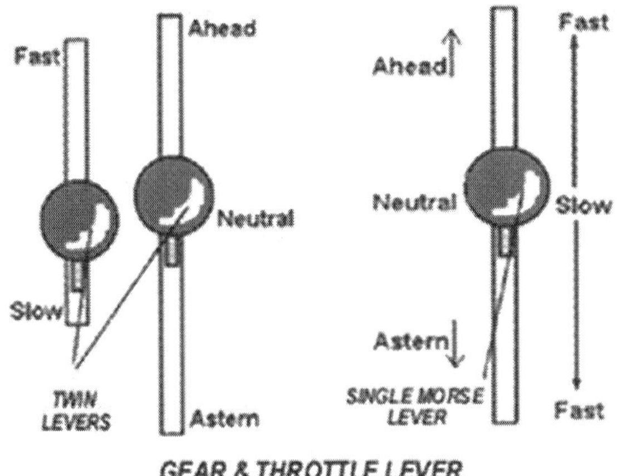

GEAR & THROTTLE LEVER ARRANGEMENTS

Modern systems operate sequentially, or have override safety systems, however good practice remains to follow the manufacturer's instructions always.

CLUTCHES

In most operations it is convenient to be able to positively engage and disengage an input shaft from an output shaft. The mechanism that smooths this operation is called a clutch.

BASIC CLUTCH MECHANISMS

Dog clutch - The simplest arrangement as found on windlasses and winches is the *dog clutch*. Shown below, a collar is splined to the driving motors shaft that can be pushed by the clutch lever to engage hard against a lug in the winch drum's side. This arrangement only allows engagement when winch and motor are stopped or both moving slowly at the same speed (a neat trick if at all possible for the arrangement shown below).

Dog clutches are not used in propulsion engines.

Cone clutch - The cone clutch does allow for engagement while the motor is running and is suitable for slow running operations such as capstans, windlasses and winches.

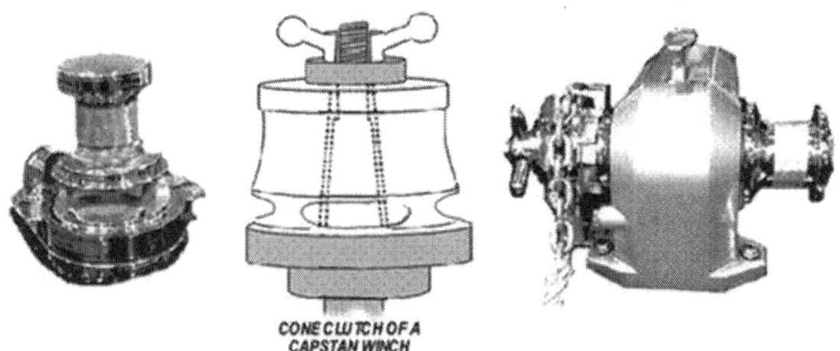

The warping drum (rope carrier) or gypsy (chain carrier) is machined with a *tapered* central hole. The drum is fitted over a matching tapered axle and can revolve smoothly when in the disengaged mode. The tapered axle can also be turned by a directly fitted motor. When the drum is tightened down onto the axle it grips (engages) and will turn with the motor driven axle. The system can snatch as it engages and although usually of heavy duty build, as with all friction type clutches there will always be momentary slip before engagement resulting in heat and wear.

More sophisticated systems are held in engagement with a heavy coil spring that can be manually compressed to disengage the drum from the axle. For a better grip the driving axle may be machined to mate inside a wide tapered drum and the mating cone's surface be coated with a friction lining material. Cone clutches are rarely used in propulsion engines.

Centrifugal clutch - A driving shaft and output coupling drum can revolve independently at their common bearing. The driving shaft has hinged weighted friction pads that at rest are restrained clear of the drum's surface by springs. When the driving shaft is revolved the friction pads are thrown outwards by centrifugal force sufficient to overcome the springs and bind onto the drums surface (engaging the coupling). Light duty pumps, power tools and the smallest outboards often use a centrifugal clutch for its light weight and economy.

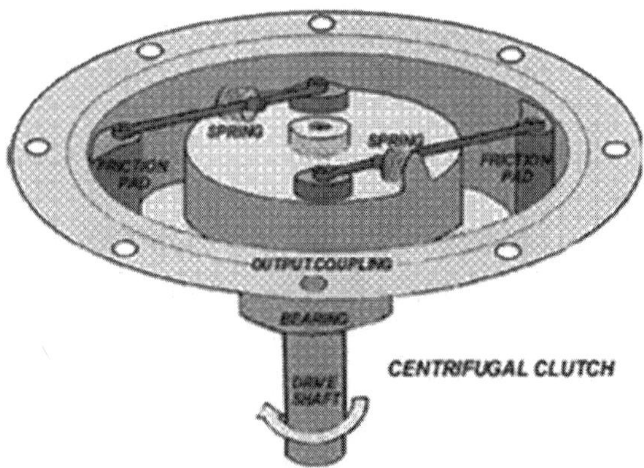

Double de-clutching - Before the modern *synchromesh gearbox* the earliest motor cars used a *crash gearbox*. To change gear, a lever is first applied to rip the currently engaged gearwheel away from the output gear train and into *neutral* (disengaging the engine from the road wheels). While temporarily in *neutral* the accelerator is revved to boost the engine speed to match the speed of the lower gear wheel (determined by the road wheel speed). This is called *double de-clutching.* When the driver judges that the speeds are matched the gear lever is applied to enmesh the lower gear by crashing it into the engine gear wheel, with hopefully not too much crunching of the gear teeth. Such foot and hand control is not available in marine gearboxes so some means of smoothing the change of gears is required.

SYNCHRONISING CLUTCH MECHANISMS

Single plate clutch - a clutch plate disc is held firmly squeezed by heavy duty springs between friction pads on the flywheel and on the pressure plate disc.

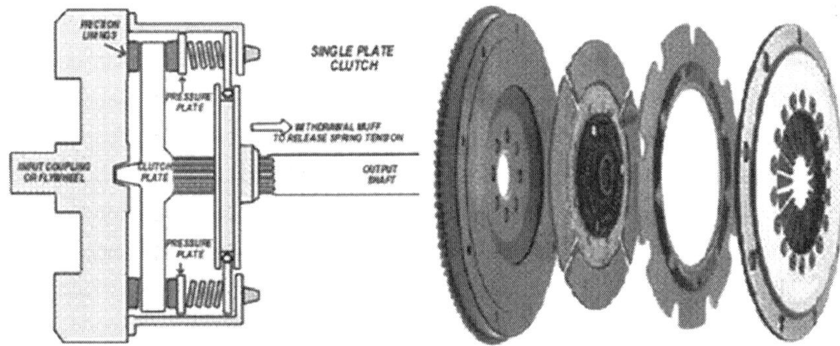

A sliding muff coupling can be withdrawn using a lever mechanism to release the spring tension and enable disengagement. Not shown in the drawing for simplicity are the photos of the flexible links between the clutch plate and the output shaft that allow for shock loads and natural oscillation in engine output.

Multiple plate clutch - The single plate clutch mechanism operates externally of the gear box and is typical for auto applications. For weight and space saving, marine applications usually have clutches that operate within the gear box. The large single clutch plate is replaced by several smaller ones enabling equivalent surface area contact in a more compact unit.

Hydraulic oil is pumped in through the input shaft which forces the internal piston disc to overcome the hold off resistance of the return spring and move towards the output assembly. This in turn squeezes together the plates that are alternately splined (embedded) inside the rim of the input shaft assembly and on the outer rim of the output shaft assembly. Adjusting the hydraulic pressure enables accommodation for heavy loading operations as may be required for low gear or reverse gear.

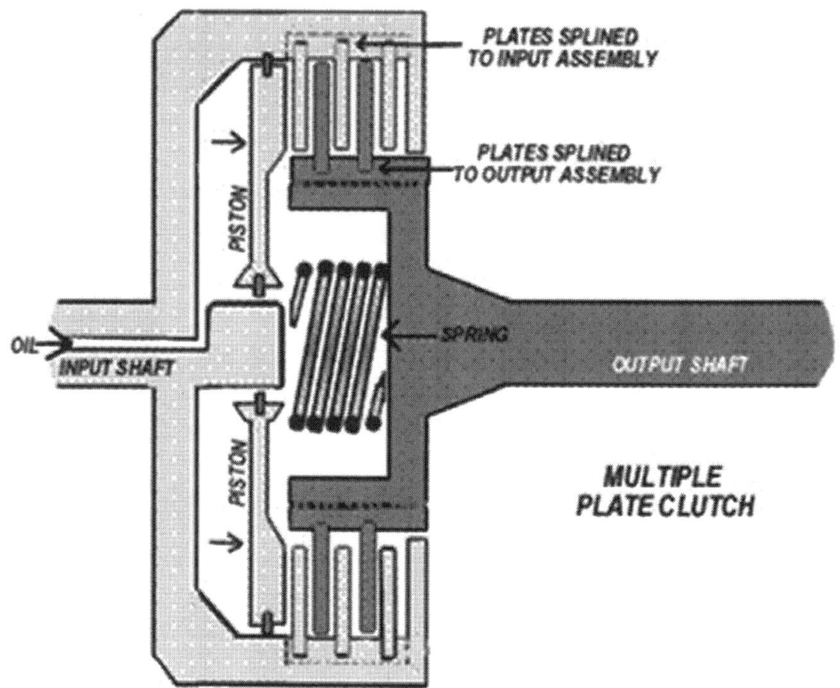

Clutch slip - Clutch slip describes when there is incomplete engagement between the driving and driven side of the clutch. The power developed by the engine is therefore not fully transmitted to the propeller. Wear of friction plates or linings is unavoidable due to momentary slippage when clutch is being engaged. Each engagement results in a slight loss of the friction material. In the course of time, this accumulated wear allows slippage. With regular maintenance, adjustments can be made to the clutch to reduce this problem but eventually wear will be such that the friction material will need to be replaced.

Fluid (hydraulic) coupling – while a hydraulic linkage may be the activating mechanism to enable positive engagement in the previous clutch types, the fluid coupling transfers thrust by smooth fluid action. Consider a household electric fan pointed towards another. The second fan will slowly rotate in the breeze. This induced motion in the driven turbine (second fan) enables a cushioned engagement/disengagement between a driving motor shaft and a driven shaft albeit with some slip (poor thrust transfer) at low speed.

A fluid coupling transmits rotation from one shaft to another by accelerating hydraulic fluid inside its housing containing closely fitted rotors; the input driving shaft pump (impeller) and the output driven shaft turbine (runner). The flywheel, impeller and housing *(shell)* are fixed and the cavity filled with hydraulic fluid. The impeller spins the fluid from the centre where the velocity is low, to the periphery where the velocity becomes high. A net force of multiplied torque on the turbine causes it to rotate with the direction of the impeller. Fluid couplings have a stall speed where the impeller is turning but not fast enough to coax the turbine into rotation (i.e. when the car driver stopped at traffic lights selects the gear while his brakes are still on). In this condition for excessive periods or if the runner becomes jammed, the engine's power could transfer its energy as overheated fluid, possibly leading to damage. A fluid coupling only achieves about 94% transmission efficiency due to fluid friction and turbulence.

Torque converter - Unlike the two rotor fluid coupling, the torque converter has at least three rotors- the impeller (motor driven), the turbine, (load driving) and positioned between them the stator (that modifies oil flow returning from the turbine to the impeller).

The fluid flow returning from a *fluid coupling's* turbine can oppose the direction of impeller rotation during slip, causing lost efficiency. The torque converter uses its stator to redirect the returning fluid to assist the rotation of the impeller so improving efficiency and output torque.

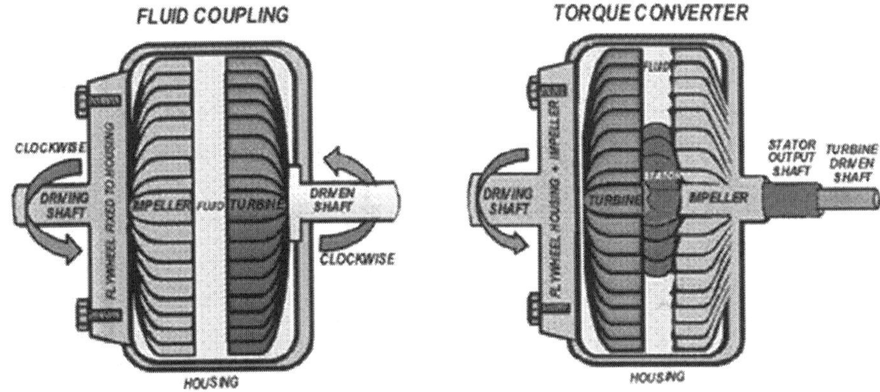

Though simple stators can be fixed often they are mounted on a one way clutch allowing forward motion only. Since the returning fluid is initially travelling in a direction opposite to impeller rotation, the stator will likewise attempt to counter-rotate as it forces the fluid to change direction, an effect prevented by the one-way stator clutch. The typical torque converter in automatic transmissions has three stages of operation:

Stall- the motor turns the impeller but the turbine cannot rotate. (When the car driver stopped at traffic lights selects the gear while the brakes are still on). Acceleration-the motor accelerates with the impeller spinning at a greater rate than the turbine.

Coupling-The turbine has reached approximately 90% of the speed of the impeller. Torque multiplication has ceased and the torque converter is behaving like a fluid coupling.

Lock-up clutches such as the Voith TurboSyn coupling overcome the problem of slip and improve fuel efficiency by coating the periphery of a multi-section impeller with friction plates. Centrifugal force throws the impeller onto the turbine at higher revolutions, creating a positive clutch lock at speed.

REVERSE AND REDUCTION GEAR BOXES

The gear system below has an input shaft and forward shaft (often combined), a reverse shaft and an output shaft turning within an oil bath called the sump (not shown). The forward and reversing shafts have pinions (driving gear wheels) engaged by clutches. When disengaged as below the reverse pinion gear wheel rotates freely on its shaft.

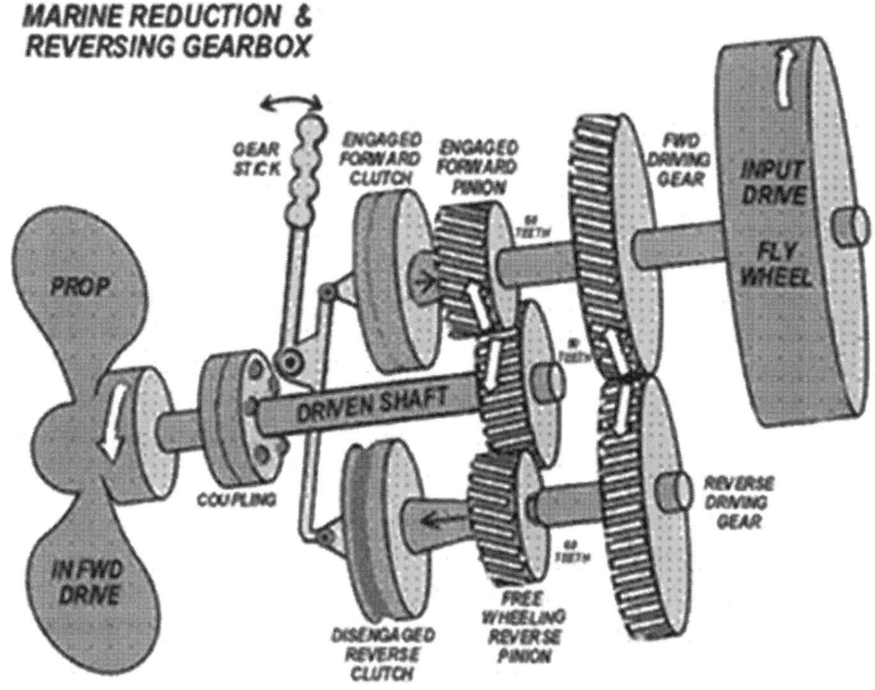

Ahead mode operation - The forward clutch pushes the forward pinion to engage on the forward/input shaft. The forward pinion then drives the driven shaft as shown above. The reverse clutch is not active so the reverse pinion rotates freely on the reverse shaft.

Astern mode operation - The reverse clutch pushes the reverse pinion to engage on the reverse shaft. The reverse pinion then drives the driven shaft as shown above. The forward clutch is not active and the forward pinion rotates freely on the forward/input shaft. An input shaft driven oil pump provides oil flow for lubrication and operation of the clutches. Oil pressure is regulated by a pressure relief valve before distribution to the gears, bearings, clutch and control via passages in the gearbox. A control box is mounted on the top of the gearbox and connected by linkage (electronic, hydraulic, pneumatic) to the throttle and forward/reverse control at the wheelhouse and gearbox (for manual emergency operation). An oil cooler and oil filters may be mounted on the gearbox.

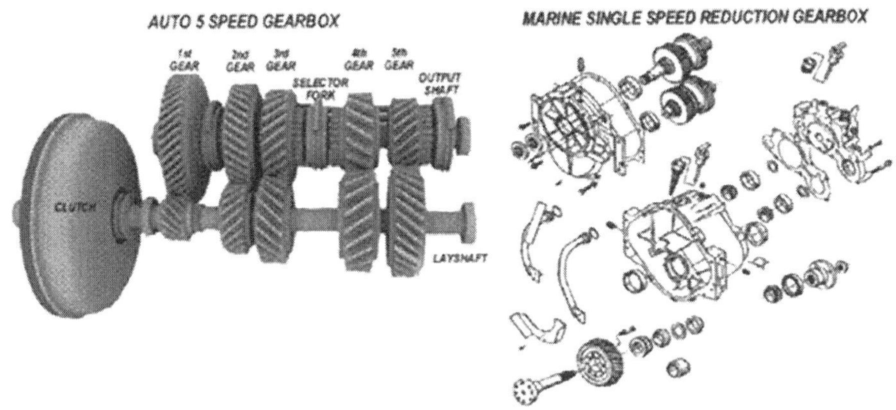

EPICYCLIC OR PLANETARY GEARING

Epicyclic or planetary gearing systems provide the advantages weigh saving, balance and compactness for a gear box suitable for marine applications. Variations of three basic geared components of a sun gear, planet gear and annulus are utilized.

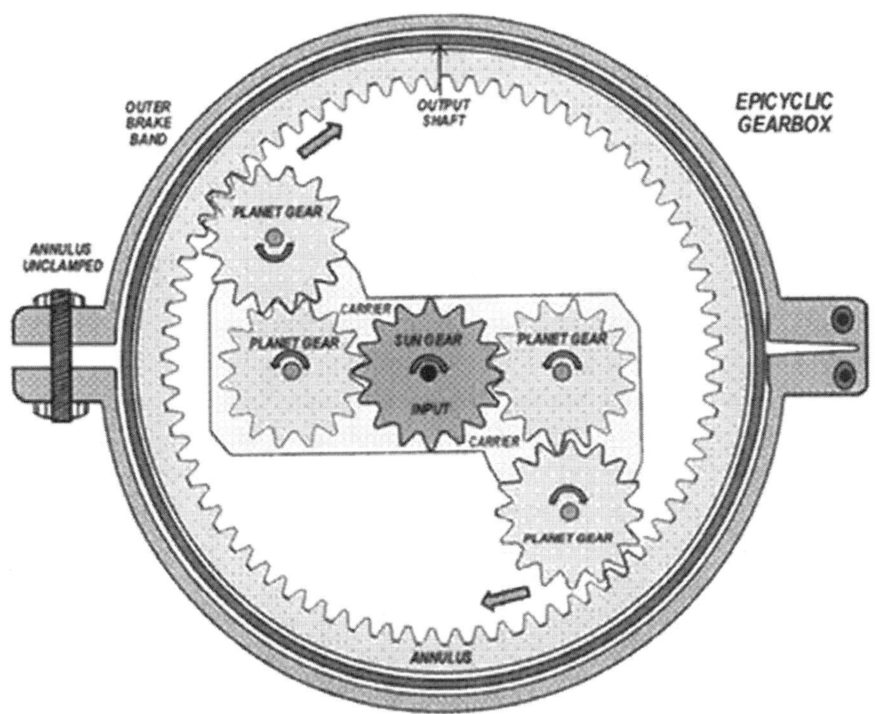

One or more outer gears (planet gears) revolve about a central gear wheel (sun gear). In more complex systems groups of planet gears can be mounted on a movable arm (carrier) which as a body can rotate around the sun gear. An outer ring (annulus) with inward-facing teeth meshes with the planet gear. In operation one of the three components (star, planet or annulus) is held stationary, another is used as a power input and the last component is used as the output. The ratio of input to output is dependent upon the number of teeth in each gear, and upon which component is held stationary.

In the reversing gearbox shown below the output shaft can revolve in ahead mode (anticlockwise) with the annulus free to rotate within the shaft. If however the annulus is locked on the output shaft by the outer tightening collar (brake band) the annulus and shaft are rotated in reverse (clockwise) by the planet gears.

Epicyclical gear arrangements can be simple or compound. Compound arrangements enable large reduction ratios, weight saving and flexibility. They may be *meshed-planet* (at least two or more planets in mesh within each planet train), *stepped-planet* (a shaft connects two planets within each planet train), or *multi-stage* structures (two or more planet sets).

6

Feed Systems in Marine Engineering

FEED WATER SYSTEM

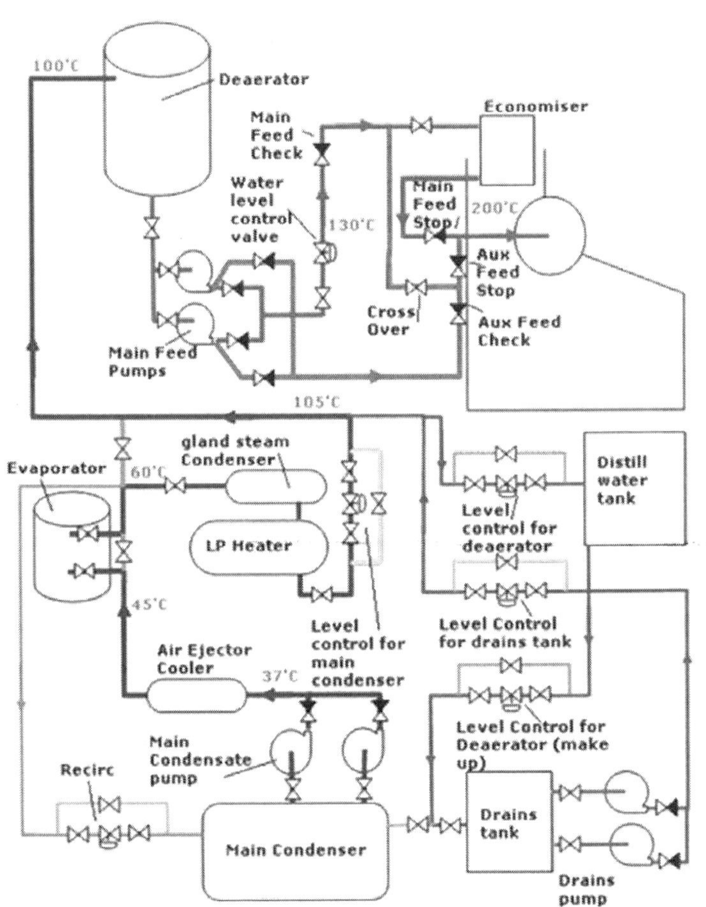

Shown above is a typical feed water system for a modern steam plant. Water is pumped from the main condenser by special centrifugal pumps

(Supercavitating) having an inducer to allow suction from the very low pressures without vaporisation of the water. The water passes through the air ejector cooler to the Condensate cooled evaporator. A recirculation valve is available to return some condensate to the main condenser. The purpose of this is to increase the overall flow though the evaporator cooler thereby increasing water production as well as to ensure a minimum flow through the Condenser extraction pumps.

The water passes via the gland steam condenser and LP heater which in this case are shown as a combined unit on to the Main Condenser level controller. This prevents the level in the condenser falling below a set level thereby causing the main condensate pumps to run dry.

Some times mounted after this is a deioniser and feed filter before the water is passed to the deaerator. The deaerator is mounted as high as possible in the engineroom increaseing the suction head for the feed pump preventing vaporisation in the suction eye of the pump. Not shown is an automatic recirculation valve fitted to the main feed pump outlet to ensure a minimum flow through the pump. The boiler water then passes via the boiler water level controller to the economiser and then through to the boiler steam drum. An second supply is available for use in emergencies to the drum either via or by passing the economiser. The drains tank condensate is pump via the drains tank level controller into the main feed system.

SYSTEM LEVEL CONTROL

Control of the amount of water in the system is by level control of the deaerator. One of the purposes of the deaerator is to act as a reserve capacity of high quality feed for the boiler. Water may be spilled to the feed tank or made up to the drains tank. An emergency filling valve is available for the main condenser the use of which is avoided as it introduces large quantities of gasses into the condenser reducing efficiency

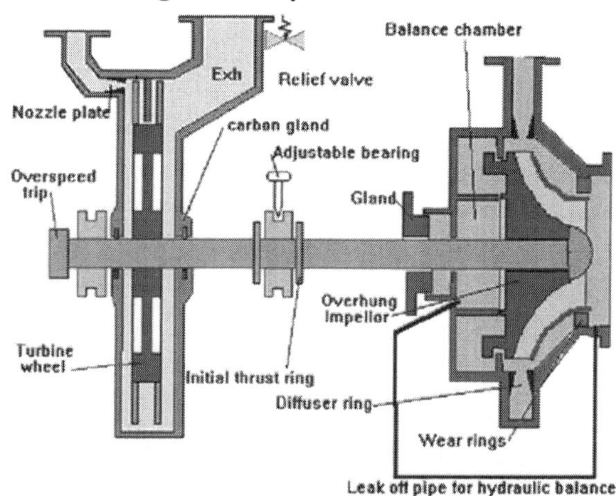

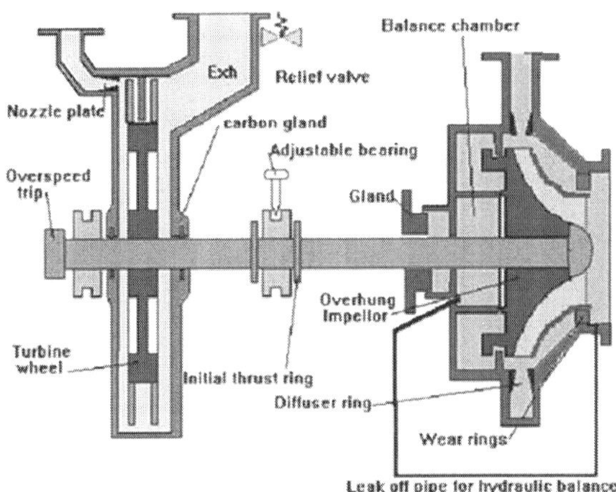

High speed, multistage centrifugal pumps are the preferred type due to their ability to supply large quantities of water and to provide it at steady flow avoiding shock loads on pipe lines and valves.

Materials

- Nozzle plate- Creep resistant steel
- Blades-Stainless iron
- Bearings-White metalled (oil lubricated)
- Turbine glands- Carbon
- Turbine casing- Cast steel
- Turbine wheel- Stainless steel
- Shaft- Nickel Chrome steel
- Wear rings- Leaded bronze
- Impeller-Stainless steel or monel
- Diffuser ring- Aluminium bronze

The impulse Curtis wheel (velocity compounded) rotates at speeds of around 7000 rpm. Velocity compounding means that there is very little pressure drop across the stages reducing the need for fine clearances. This allows the turbine to be run up quickly from cold and with its inherent strength allows for rapid speed/load variations

The balance chamber must have a diameter greater than the suction wear ring. The high pressure on the back side of the impeller would produce an axial loading pushing the impeller towards the low pressure of the suction eye and overloading the bearings.

Bleeding away the high pressure on the back side of the impeller removes this effect but looses pump efficiency through recirculation. The balanced piston design allows for the bleeding but restricts the flow to only that necessary.

As the pressure builds up the impeller is pushed towards the suction eye which increase the leakage into the balance chamber reducing the pressure at the back of the impeller. The high pressure on the forward face pushes the impeller back away from the suction eye and so the leakage to the balance chamber reduces and the pressure increases on the back face of the impeller

If the pump is designed to be supercavitating hard metal inducers are fitted which screw into the water. Any cavitation occurs on this which is made sacrificial. Supercavitating pumps are required where the feed water temperature is close to the saturation temperature so any pressure drop causes the water to flash off to steam Inlet steam pressure around 60 bar, outlet around 3 bar. Expansion down to lower pressures would require excessively large casings and would lead to problems of centrifugal stresses due to the larger blading required. Carbon seals are used instead of labyrinths for simplicity and to keep the length of the unit down to a minimum. Due to the different coefficients of expansions between the carbon and the steel a madrill must be used to set the correct running clearance. For multistage pumps an extra end bearing is required and hence additional packed gland.

The pump is dynamically balanced by means of the balance chamber leak off to the suction eye, and the dam edges on the back of the impeller

BALANCING OF TWO STAGE PUMPS

For two stage pumps it is possible to dynamically balance then by having the impellers back to back. More typically the balance piston assembly is fitted to the final stage Feed pumps of this type of balance are best started against either a closed or spring loaded discharge v/v to ensure rapid build up of pressure

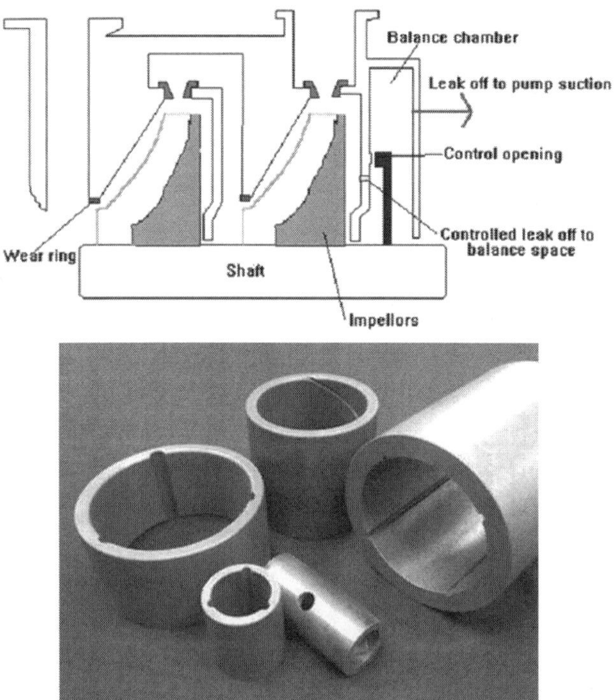

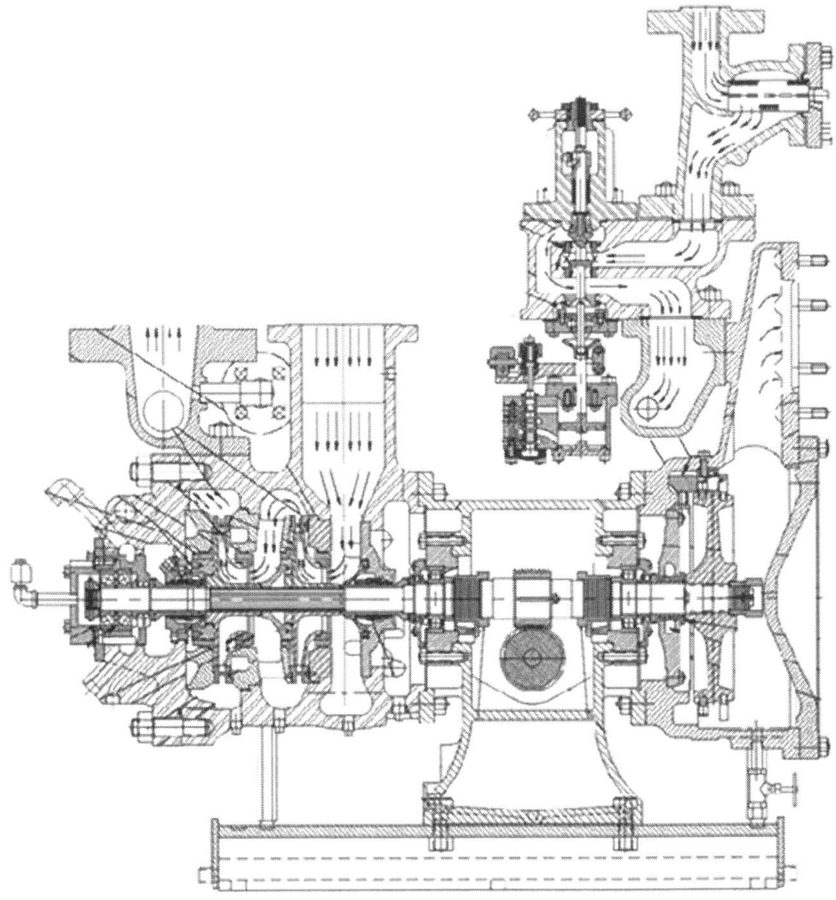

PUMP SIDE BEARINGS

Water lubricated bearings- Steel backing onto which is sintered a layer of porous bronze impregnated with PTFE (0.025mm thick). This PTFE is transferred to the shaft so providing very low coefficients of friction

Bearings operate at 115oC with water supplied at 5.5bar 70oC.

Bearing clearance- 0.15 mm

Max- 0.25mm must be replaced

Danger- 0.3 mm severe damage will occur

The bearings should also be replaced if less than 75 % shows on the surface.

The coffin feed pump balances the axial load of the pump end with the axial load of the turbine. The excess is absorbed in the thruster arrangement.

Inexplicably in his early career the author was let loose on one of these and can attest to the superb design including the arrangement of tooling for disassembly/assembly. Unfortunately the mix of left and right hand threads caused 'issues'!

COFFIN FEED PUMPS

FLOW CHARACTERISTICS OF CONTROLLED FEED PUMPS

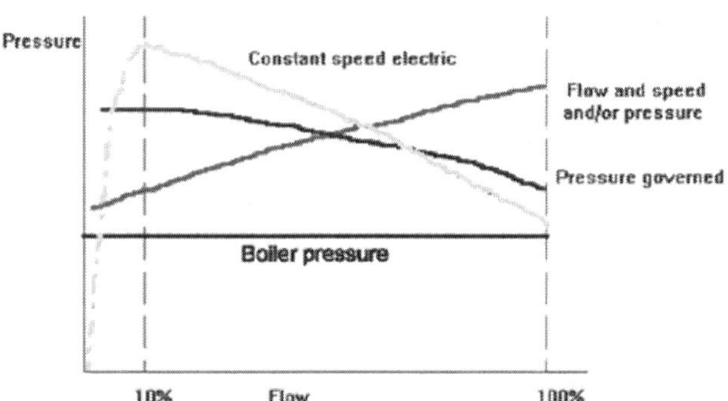

Constant speed (electric)- Below 10% turbulence becomes so great that the pump operates very inefficiently and must not be worked in this range. Hence, a recirc valve must be used. At full flow pressure is only just sufficient to feed the boiler

Pressure governed- Pressure droop designed in to give stable control.

Flow + Speed and/or Pressure- The extra element must be added with the flow otherwise the system becomes unstable. Characteristic can be level or slightly rising giving low pressure at low flow rates conserving energy and preventing water being forced at high pressures through partly open feed control valve.

GLAND

Can be operated at very low flow rates due to reduced speed

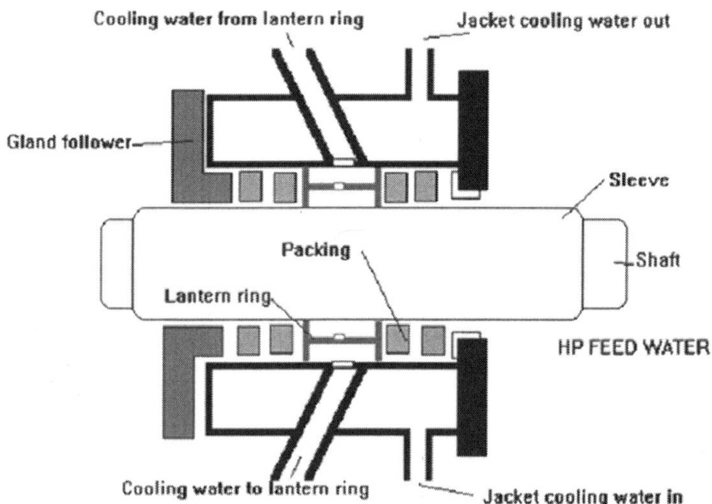

The purpose of feed heating is to increase plant efficiency. With no feed heating steam gives up three times more heat to the cooling water as it does in doing useful work in the turbine. This major loss occurs in Condenser with the low quality exhaust steam being condensed to enable it to be returned to the boiler/

As the steam expands through the turbines so its specific energy content falls and the volume increases.

A proportion of the low quality steam is bled off from the turbines to a feed heater where it can give its energy (its latent heat mostly) to the feed and so a higher proportion of the energy is reclaimed by the system as would have been gained by expanding it through the turbine to the condenser.

Feed heating can take two forms:

- Contact heating, such as within the Deaerator where the steam and feed water mix and and the water directly receives heat.
- Indirect heating, such as LP Feed heaters where the condensate and heating steam are separated for example with the condensate passing through the steam environment within tubes

CONDENSATE FEED HEARING

Theoretical practical cycle where a portion of the steam is bled off in stages and used for feed heating

It should be noted that although thermal efficiency increases with the number of stages, it is governed by the law of diminishing returns and the improvement is reduces with successive stages. Hence the cost of the increased plant becomes a factor.

Bleeding off a portion of the steam gives the added advantage that the steam volume that has to be accommodated in the final stages of the turbine and condenser is reduced.

Theoretical cycle with all steam expanded through the turbine

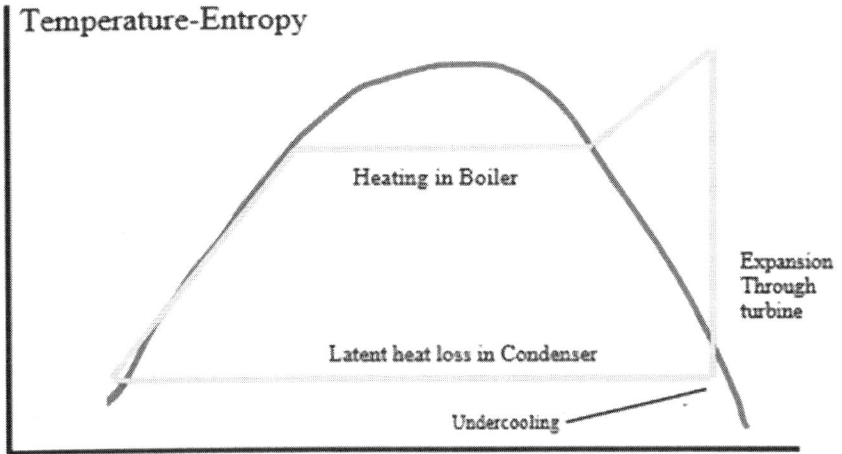

Theoretical cycle where all the steam is continuously used in Feed Heating 100% efficiency

This could only be possible where the turbines where fitted with a water jacket through which the feed water flows.

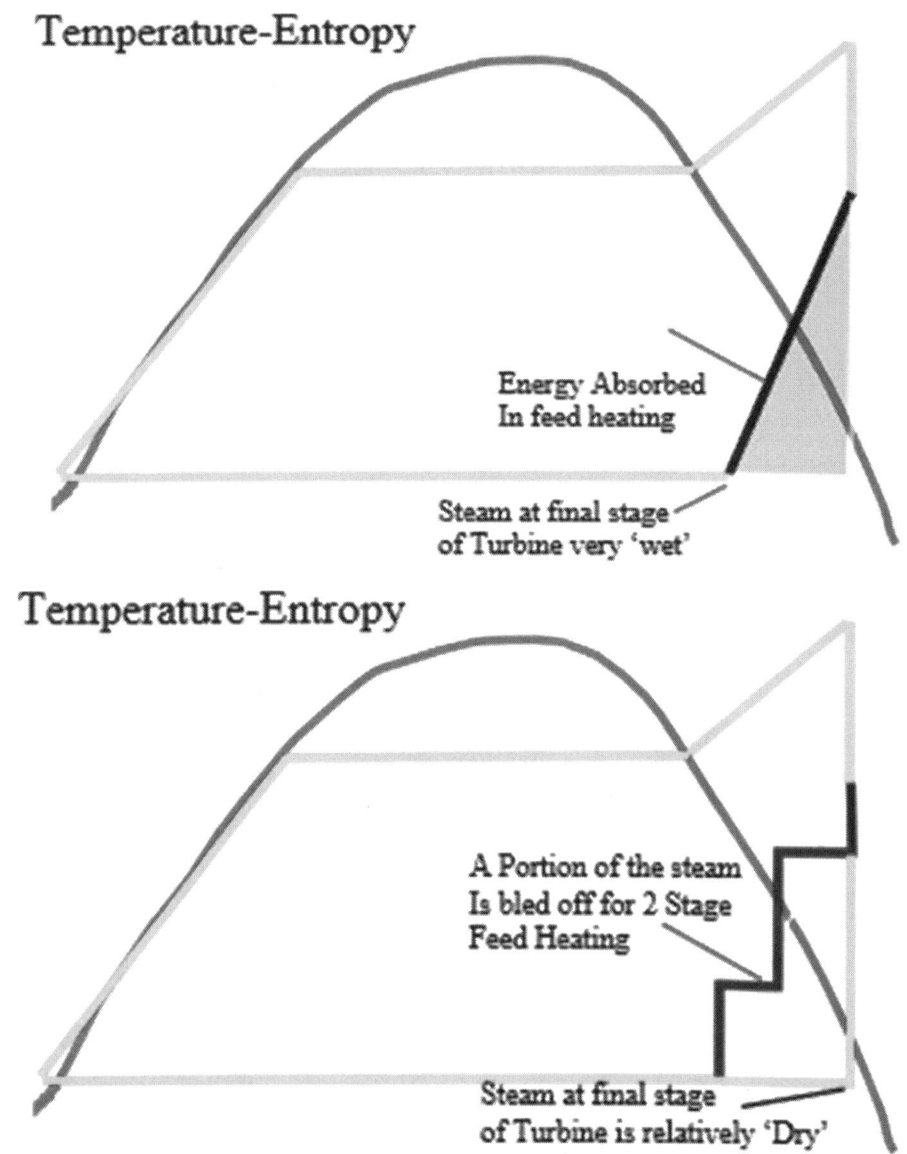

HP FEED HEATER

HP feed heater as fitted to cascade feed heater system instead of economiser

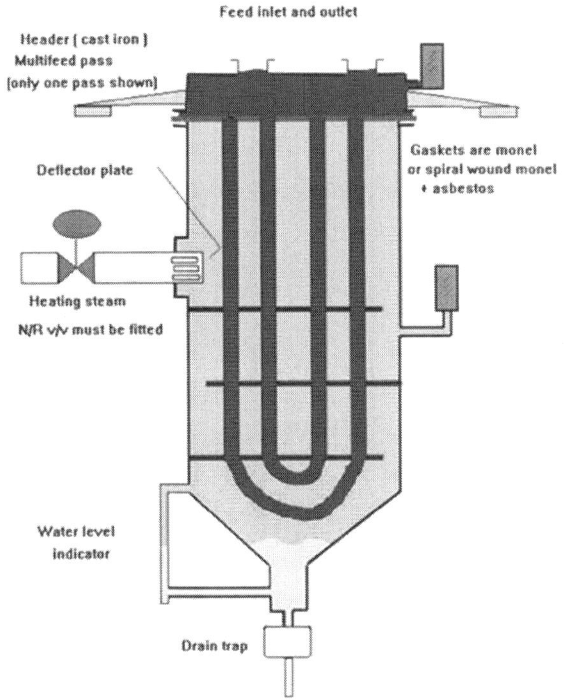
Feed inlet and outlet
Header (cast iron)
Multifeed pass
(only one pass shown)
Gaskets are monel
or spiral wound monel
+ asbestos
Deflector plate
Heating steam
N/R v/v must be fitted
Water level
indicator
Drain trap

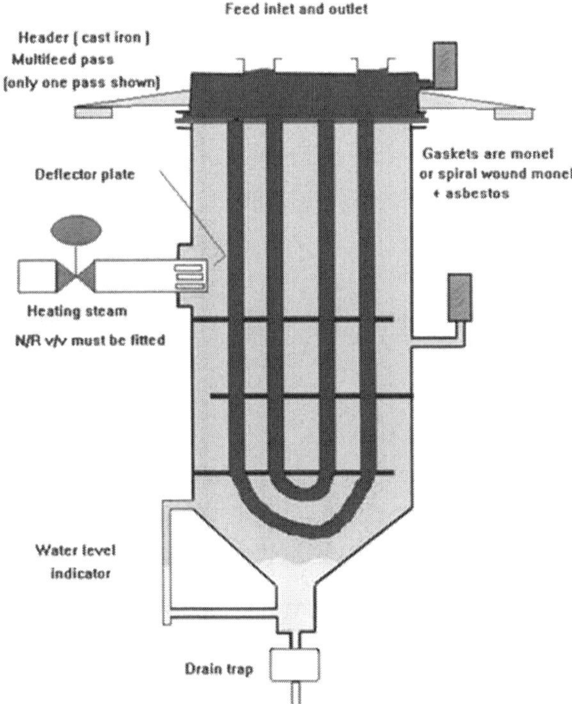
Feed inlet and outlet
Header (cast iron)
Multifeed pass
(only one pass shown)
Gaskets are monel
or spiral wound monel
+ asbestos
Deflector plate
Heating steam
N/R v/v must be fitted
Water level
indicator
Drain trap

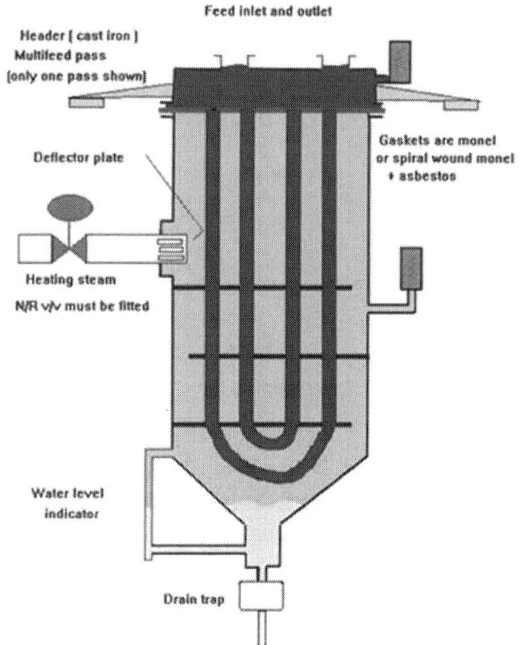
Feed inlet and outlet
Header (cast iron)
Multifeed pass
(only one pass shown)
Deflector plate
Gaskets are monel
or spiral wound monel
+ asbestos
Heating steam
N/R v/v must be fitted
Water level
indicator
Drain trap

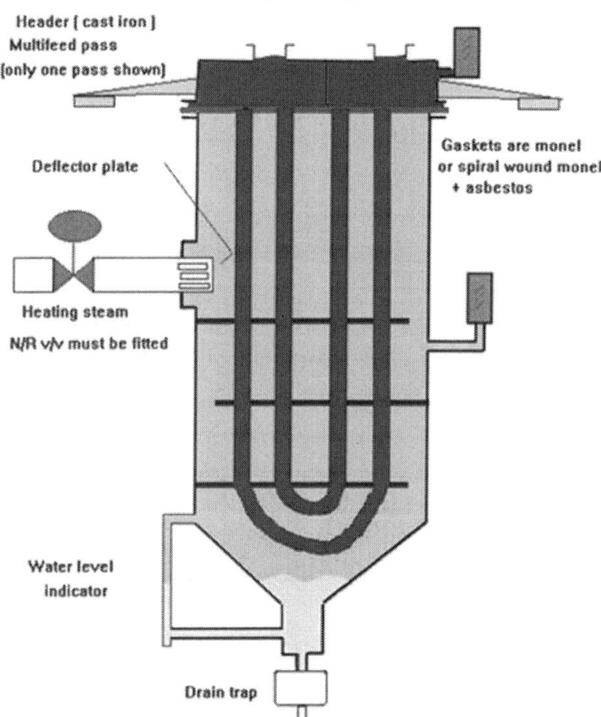
Feed inlet and outlet
Header (cast iron)
Multifeed pass
(only one pass shown)
Deflector plate
Gaskets are monel
or spiral wound monel
+ asbestos
Heating steam
N/R v/v must be fitted
Water level
indicator
Drain trap

- The high pressure tubes are made of copper nickel or steel.
- The casing is made of mild steel

For efficiency it is important that the bottom of the heater does not become flooded.

As a large difference in pressure can be accommodated between the feed and steam, this type of heater can be fitted on the discharge side of the main feed pumps; in a cascade feed heating system this replaces the economiser and the heat in the flue gas is recovered by a regenerative air heater (Lungstrom). This system allows the feed to be heated to a high temperature

The change in temperature is normally about 30oC

The heater, when new, must be able to withstand either or both of the following;

- 2 x boiler press + 20%
- 125 % feed press continuously

ECONOMISERS

PURPOSE

The purpose of the economiser is to increase plant efficiency by removing heat that would otherwise be lost in the flue gas and use it to indirectly feed heat the water. By heating the feed water it is also helping to prevent thermal shock as the water enters the steam drum.

DESCRIPTION

The flow of water is general counterflow. The exception to this is in the radiant heat boiler where the economiser is mounted immediately above the superheaters. The water flow in these so called 'Steaming economisers' is then parallel flow to decrease the tendency for the economiser to steam excessively above the design limits and lead to steam blockage, this is why these economisers are bare tube with no extended heating surface.

Vent and drains are fitted to header where isolating valves are fitted a safety valve must also be added.

The modern design involves the fabrication of the header and stub tubes which can then be heat treated. The tubes are then site welded to the stubs.

The previous use of expanded joints has now fallen out of failure due to the requirement of a multitude of hand hole doors with associated joints and hence possible area of leakage. The materials used are governed the materials susceptibility to cold end corrosion, any metal having a surface temperature below the dew point will tend to have acidic deposits forming on its surface caused by the water absorbing sulphur trioxide and dioxide from the flue gases. Some metals are more resistant to this form of corrosion at lower temperatures, choice of material will initially depend on the minimum metal surface temperature and is calculated as the feed water temperature that is passing through the economiser plus 5oC. For temperatures greater than 138oC solid drawn mild steel tubing is used. Fitted to these are welded on extended surface steel fins or studs.

For temperatures between 115°C to 138°C shrunk on or cast iron gills must be used. The temperature should not be allowed to fall below 115oC as this can lead to heavy fouling as well as corrosive attack.

Efficient sootblowing is absolutely essential to ensure that surface are kept clear of combustion products which can not only lead to heavy corrosion and a drop in efficiency, but also to the possibility of an economiser fire with potentially disastrous consequences. With this in mind it is not unusual to find provision for water washing, something which is carried out on a very regular basis on a Motor ship with waste heat recovery.

If due to failure it is required to run the economiser dry then the maximum gas inlet temperature should be limited to about 370oC, vents and drains should be left open to ensure that there is no build up of pressure from any water that may be still located in the tubes.

CAST IRON EXTENDED GILL ECONOMISER

INLET END

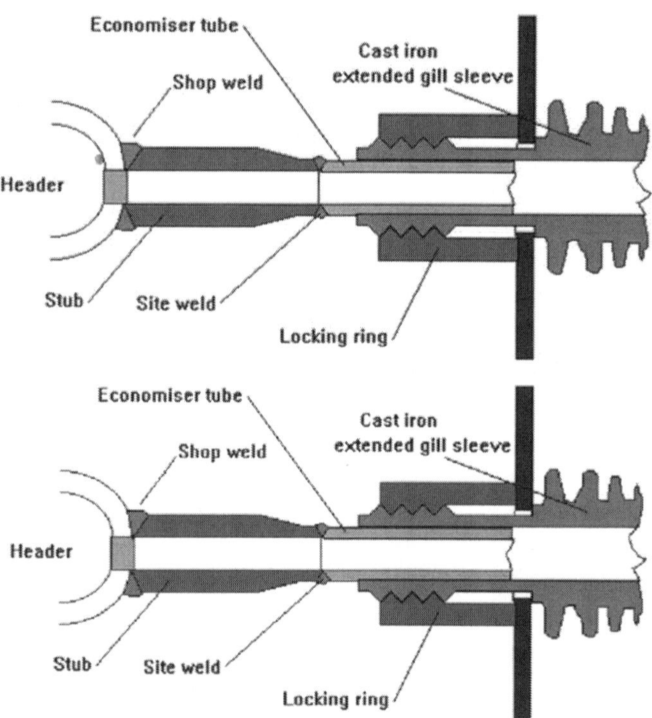

OUTLET END

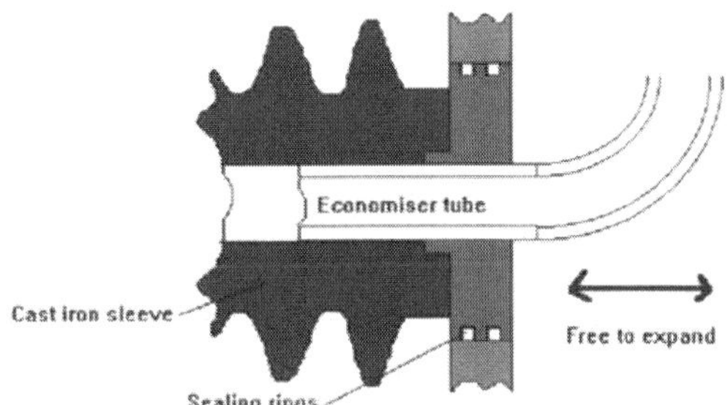

The unit is of the multiple pass 'melesco' style, the sleeves may be slid on before the bends are welded on.

The header is made out of mild steel.

MILD STEEL FIN EXTENDED SURFACE ECONOMISER

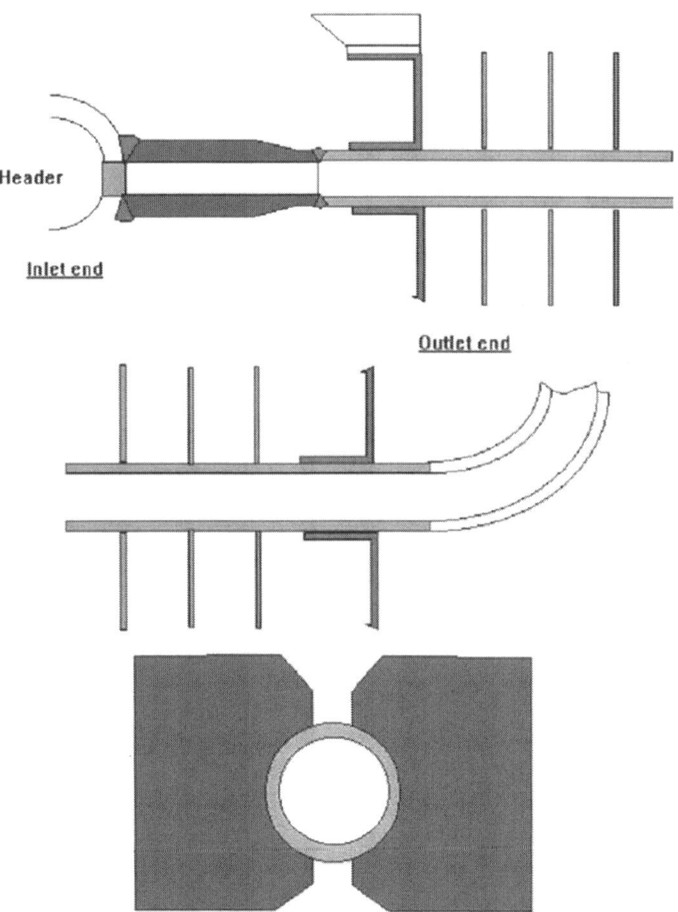

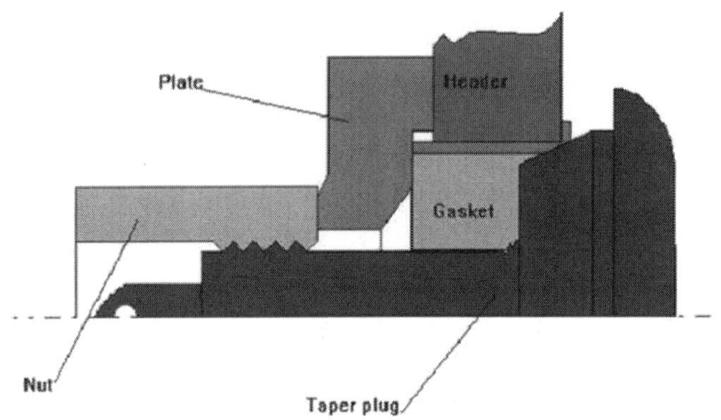

DEAERATOR

In marine water tube boilers it is essential to keep water free of dissolved gases and impurities to prevent serious damage occurring in the boiler

In a closed feed system the regenerative condenser removes the bulk of the gases with a dissolved oxygen content of less than 0.02 ml/l

However, it is recommended for boilers operating above 30 bar and essential for those operating above 42 bar that a dearator be fitted.

PURPOSE OF FITTING A DEAERATOR

There are four main purposes;

- To act as a storage tank so as to maintain a level of water in the system
- To keep a constant head on the feed system and in particularly the Feed pumps.
- Allow for mechanical deaeration of the water
- Act as a contact feed heater.

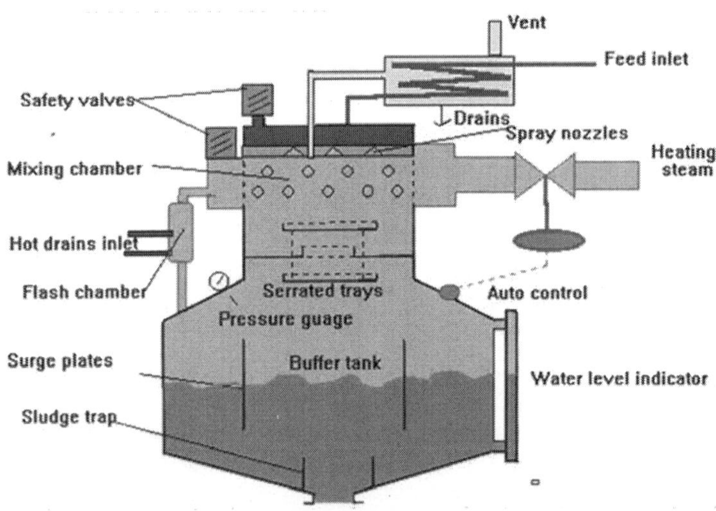

Feed water enters the dearator via the vent cooler, here the non condensible gasses and a small amount of steam vapour are cooled. The condensed water is returned to the system. The feed water is sprayed into the mixing chamber via nozzles, for systems with large variations of flow, two separate nozzle boxes may fitted with two independent shut off valves to ensure sufficient pressure for efficient spray.

Alternatively, automatic spray valves may be fitted.

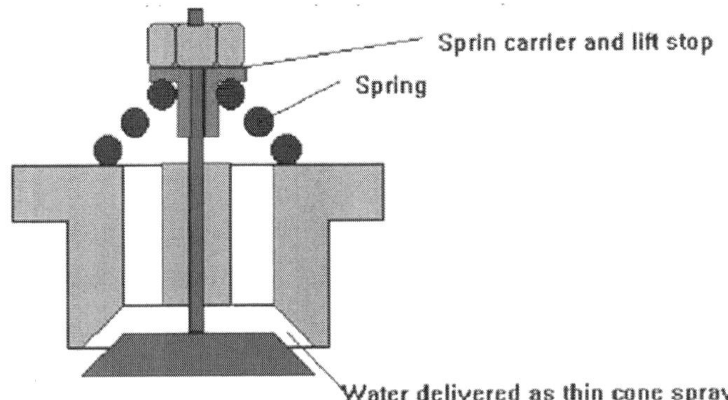

By varying the spring tension the pressure at which the nozzles open can be set at different levels.

The water in a fine spray, and so high surface area to volume, is rapidly heated to corresponding saturation temperature in the mixing chamber by the heating steam. This steam is supplied from the exhaust or IP system and is at about 2.5 bar.

The heated water and condensed steam then falls onto a series of plates with serrated edges, the purpose of these is to mechanically remove any gas bubbles in the water improving the efficiency of the process.

Finally the water falls into the buffer tank, before exiting to the feed pumps; as the water is now at saturation temperature any drop in pressure (such as in the suction eye of the impeller) will cause vapour bubbles to form. Hence, the deaerator must be fitted well above the feed pumps or alternatively an extraction pump must be fitted to supply the feed pumps.

Hot water drains are led to the dearator where they are allowed to flash into steam adding to the heating steam.

The non condensible gas outlet is limited so that there is only a small flow of water from the drain of the cooler. This water should be discarded as it contains not only high quantities of oxygen but also ammonia.

Requirements for efficient operation

- The minimum temperature increase is 28oC
- The minimum heating steam temperature is 115oC which corresponds to a pressure of 1.7 bar; this is to prevent deposits forming on the economiser. To prevent cold end corrosion on cast iron protected

economisers this temperature rises leading to the common practice of using steam at 2.5 bar

- The water at inlet should have an oxygen content of not more than 0.02 ml/l

With these parameters met the water at outlet should have an oxygen level of not more than 0.005 ml/l.

A thermometer is fitted to the shell, the temperature should be kept to within 4oC of the saturation temperature of the pressure indicated on the pressure gauge. This can be governed by the quantity of heating steam added. If the difference is always greater the partial pressure due to the non-condensable gasses is high and the possibility of redissolving increases. Where fitted the valve on the air outlet should be opened more to limit of too much steam being lost up the vent. The fitting of vent condensers is not universal and it is not uncommon to see the vent led up the funnel where a small wisp of steam can be seen when correctly adjusted.

FEED TANKS

A well designed feed tank should be designed to minimise the oxygen within the feed system. This is especially important with open feed systems.

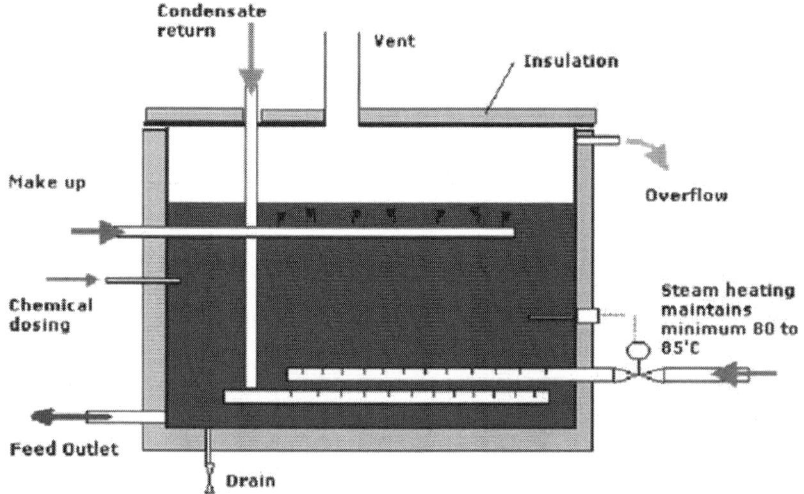

The following are taken as parameters for a well designed tank

- Adequate ventilation with on or more vent pies determined by the volume of water
- Condensate enters as low as possible via a slotted sparge pipe
- Cold water make up enters at highest point
- Sufficient tank volume to cope with transient flows from normal operations without necessitating spilling back to feed tanks or overflows.

- Tank to have sufficient volume of water at normal working level to allow for 1 hours operation at maximum demand.
- Take off to feed pumps to be at least 75mm from tank bottom
- Tank to be located to provide head requirements at normal working level for feed pumps

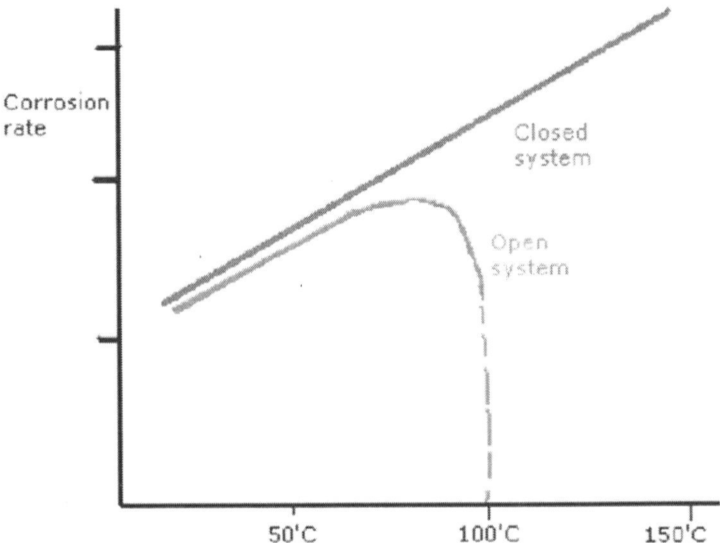

As a rule of thumb chemical reaction rates double for evey 10'C rise in temperature. For feed system this remains true upto about 80'C for open system. After this due to reducing solubility of oxygen the rate of corrosion reduces. Thus steam heating on the open feed tanks have thermostats set at 85'C or higher

STEAM - STEAM GENERATORS

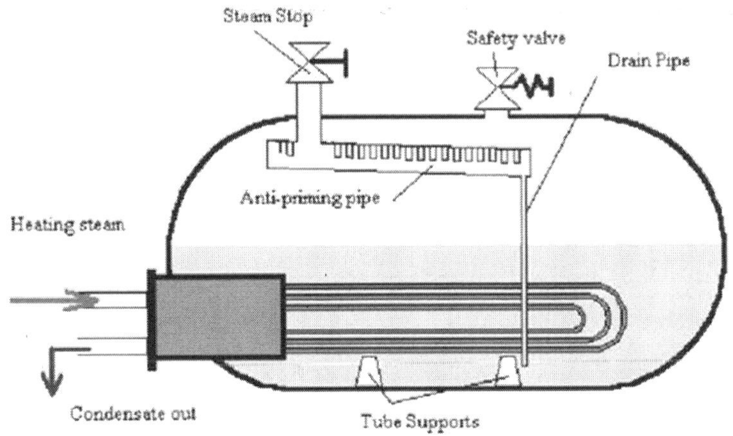

Found on steam propulsion plant and used for the production of low pressure steam for tank heating purposes. This has the major advantage in reducing the possibility of damage due to oil contamination where the feed system is segregated.

Generally the heating steam is supplied from the Intermediate Pressure system when under way, additional 'Live' steam make up may be required under low system load conditions.

CLOSED FEED SYSTEM?

The main difference between an open feed system and a closed feed system is that in a closed feed system none of the parts of the system are exposed to the environment, as you will later see in the feed water system diagram.Closed feed system is generally used in a high pressurized system such as a high pressure water tube boiler supplying steam to themain propulsion steam turbine.

The main parts of a closed feed system are:

- A regenerative type condenser
- Air ejector
- Extraction pump
- De-aerator
- Feed tank.

THE CLOSED FEED SYSTEM

The high pressurized steam from the boiler enters the steam turbine and exits in the form of low energy steam, which then enters the condenser. A closed feed system mostly consists of a regenerative type of condenser, which allows condensation of the steam with less drop in the temperature level. Once the condensate collects in the condenser, it is extracted with the help of extraction pumps and is then circulated to the air ejector.

The air ejector does the work of removing dissolved oxygen from the water. This helps to preventcorrosion of internal parts of the boiler. Thecondensate is heated as it passes through the condenser. The air is removed from the condensateusing steam operated ejectors.

After removal of the oxygen, the condensate is passed through a gland steam condenser, which further heats the condensate. The gland steam condenser is a heat exchanger in which the turbine gland steam is condensed and drained to the atmospheric drain tank. The condensate after getting heated is then passed through a low pressureheater, which uses bled steam from the turbine.

The main aim of using all these condensers and heatexchangers is to prevent wastage of heat from the system and thus increase the feed water temperature to assist in the de-aeration process.

DE-AERATOR

The de-aerator used in a closed feed system is a direct contact feed heater. This means that in the heat transfer process, the steam and feed water literally comes in direct contact with each other and actually mixes. As the temperature of the water rises, any dissolved oxygen present in the feed water is released.

There is a feed water storage tank at the bottom of the de aerator, which is used to supply feed water to the main feed pumps. One of the two feed pumps supplies the required feed water to the boiler. The feed water before reaching the boiler, passes through a high pressure feed heater. The high pressure feed heater further increases the temperature of the feed water. The feed water then passes through an economizer, before finally entering the boiler drum.

The atmospheric drain tank of the system collects drains from various installations such as gland steam, air ejectorsteam etc. The main need of the atmospheric drain tank is to supply additional feed water to the system whenever needed. In order to prevent any kind of mishap during low load and manoeuvring operations, a recirculating feed line is provided to ensure adequate flow of feed water through the air ejector and gland steam condenser.

According to the type of the ship and the volume of steam required, various modification are made in the closed feed system.

OPEN FEED SYSTEM FOR AN AUXILIARY BOILER - HOW IT WORKS

OPEN FEED SYSTEM

The feed system completes the cycle between boiler and turbine to enable the exhausted steam to return to the boiler as feedwater. The feed system is made up of four basic items: the boiler, the turbine, the condenser and the feed pump. The boiler produces steam which is supplied to the turbine and finally exhausted as low-energy steam to the condenser. The condenser condenses the steam to water (condensate) which is then pumped into the boiler by the feed pump.

OPEN FEED SYSTEM

An open feed system for an auxiliary boiler is shown in Figure. The exhaust steam from the various services is condensed in the condenser. The condenser is circulated by sea water and may operate at atmospheric pressure or under a small amount of vacuum. The condensate then drains under the action of gravity to the hotwell and feed filter tank.

Where the condenser is under an amount of vacuum, extraction pumps will be used to transfer the condensate to the hotwell. The hotwell will also receive drains from possibly contaminated systems, e.g. fuel oil heating system, oil tank heating, etc. These may arrive from a drains cooler or from an observation tank. An observation tank, where fitted, permits inspection of the drains and their discharge to the oily bilge if contaminated. The feed filter and hotwell tank is arranged with internal baffles to bring about preliminary oil separation from any contaminated feed or drains. The feedwater is then passed through charcoal or cloth filters to complete the cleaning process. Any overflow from the hotwell passes to the feedwater tank which provides additional feedwater to the system when required. The hotwell provides feedwater to the main and auxiliary feed pump suctions. A feed heater may be fitted into the main feed line.

This heater may be of the surface type, providing only heating, or may be of the direct contact type which will de-aerate in addition. De-aeration is the removal of oxygen in feedwater which can cause corrosion problems in the boiler. A feed regulator will control the feedwater input to the boiler and maintain the correct water level in the drum.

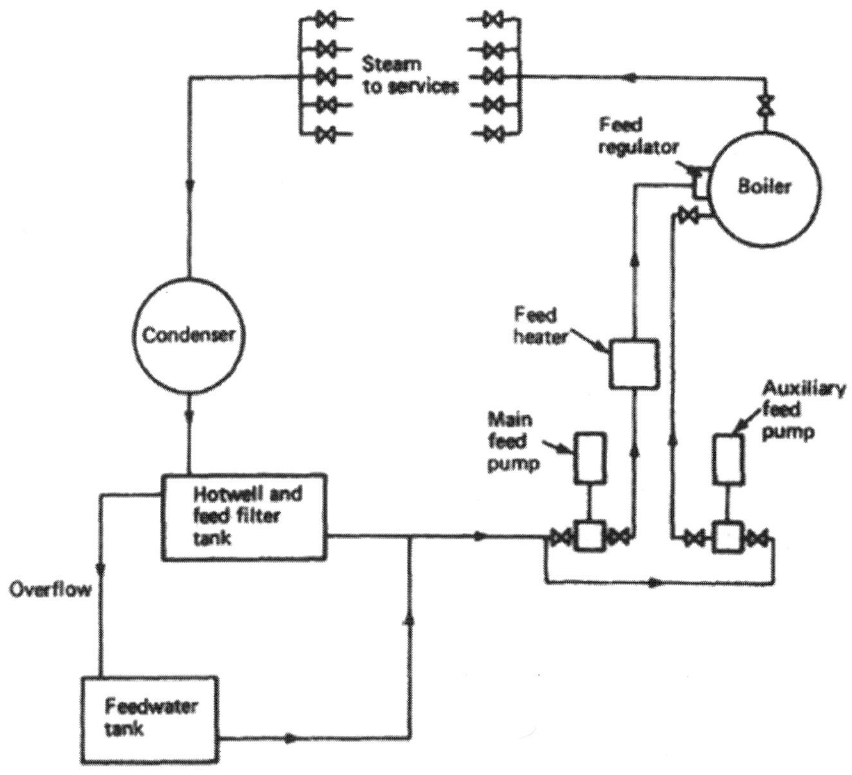

Fig. Energy conversion in a Open feed system

The system described above can only be said to be typical and numerous variations will no doubt be found, depending upon particular plant requirements.

STEAM TURBINE

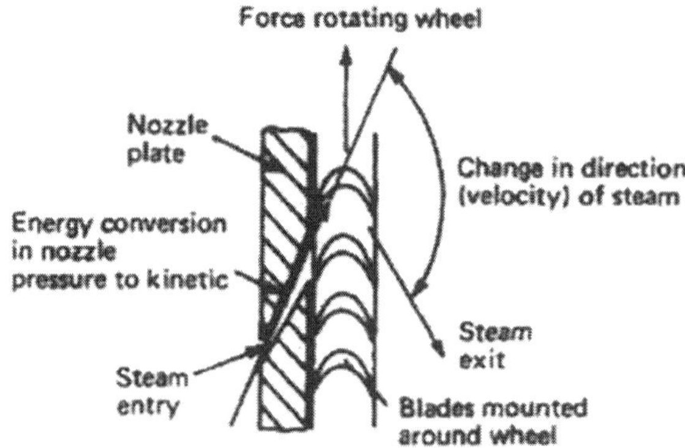

Fig. Energy conversion in a steam turbine

The steam turbine is a device for obtaining mechanical work from the energy stored in steam. Steam enters the turbine with a high energy content and leaves after giving up most of it. The high-pressure steam from the boiler is expanded in nozzles to create a high-velocity jet of steam. The nozzle acts to convert heat energy in the steam into kinetic energy. This jet is directed into blades mounted on the periphery of a wheel or disc.

CLOSED FEED SYSTEM FOR AN AUXILIARY MARINE BOILER & OPERATING PRINCIPLE

HOW A CLOSED FEED SYSTEM WORKS ?

The feed system completes the cycle between boiler and turbine to enable the exhausted steam to return to the boiler as feedwater. The feed system is made up of four basic items: the boiler, the turbine, the condenser and the feed pump. The boiler produces steam which is supplied to the turbine and finally exhausted as low-energy steam to the condenser. The condenser condenses the steam to water (condensate) which is then pumped into the boiler by the feed pump.

CLOSED FEED SYSTEM

A closed feed system for a high pressure watertube boiler supplying a main propulsion steam turbine. The steam turbine will exhaust into the condenser which will be at a high vacuum.

A regenerative type of condenser will be used which allows condensing of the steam with the minimum drop in temperature. The condensate is removed by an extraction pump and circulates through an air ejector.

The condensate is heated in passing through the air ejector. The ejector removes air from the condenser using steam-ope rated ejectors. The condensate is now circulated through a gland steam condenser where it is further heated. In this heat exchanger the turbine gland steam is condensed and drains to the atmospheric drain tank. The condensate is now passed through a low-pressure heater which is supplied with bled steam from the turbine. All these various heat exchangers improve the plant efficiency by recovering heat, and the increased feedwater temperature assists in the de-aeration process. The de-aerator is a direct contact feed heater, i.e. the feedwater and the heating steam actually mix.

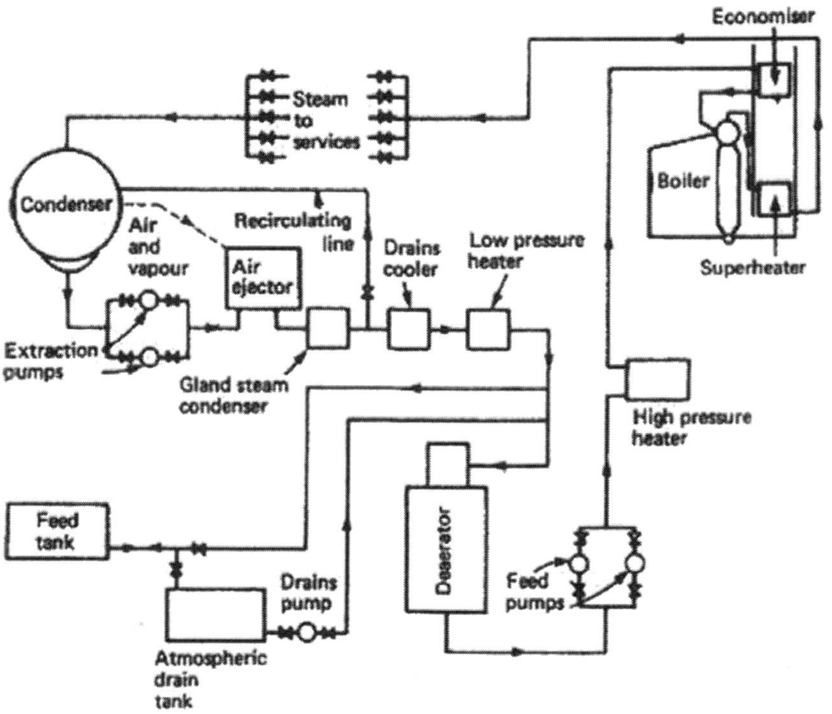

Fig. Energy conversion in a Closed feed system

In addition to heating, any dissolved gases, particularly oxygen, are released from the feedwater. The lower part of the de-aerator is a storage tank which supplies feedwater to the main feed pumps, one of which will supply the boiler's requirements. The feedwater passes to a high-pressure feed heater and then to the economiser and the boiler water drum. An atmospheric drain tank and a feed tank are present in the system to store surplus feedwater and supply it when required. The drain tank collects the many drains in the system such as gland steam, air ejector steam, etc. A recircuiating feed line is provided for low load and manoeuvring operation to ensure an adequate flow of feedwater through the air ejector and gland steam condenser.

The system described is only typical and variations to meet particular conditions will no doubt be found.

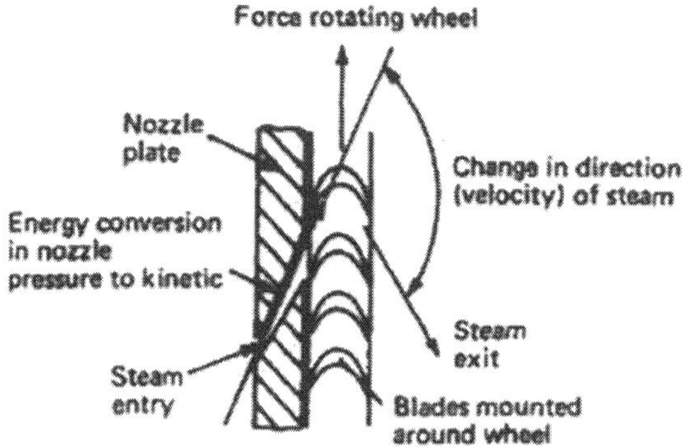

Fig. Energy conversion in a steam turbine

WHAT IS A STEAM TURBINE

The steam turbine is a device for obtaining mechanical work from the energy stored in steam. Steam enters the turbine with a high energy content and leaves after giving up most of it. The high-pressure steam from the boiler is expanded in nozzles to create a high-velocity jet of steam. The nozzle acts to convert heat energy in the steam into kinetic energy. This jet is directed into blades mounted on the periphery of a wheel or disc.

BOILER FEED SYSTEM OPERATION AND MAINTENANCE PROCEDURE

BALANCING BOILER FEED INPUT AND STEAM OUTPUT

During operation the feed system must maintain a balance between feed input and steam output, together with a normal water level in the boiler. The condenser sea water boxes are protected by sacrificial mild steel plates which must be renewed regularly. The tube plates should be examined at the same time to ensure no erosion has taken place as a result of too high a circulating water speed. Any leaking tubes will cause feedwater contamination, and where this is suspected the condenser must be tested.

Extraction pumps should be checked regularly to ensure that the sealing arrangements are preventing air from entering the system. It is usual with most types of glands to permit a slight leakage of water to ensure lubrication of the shaft and the gland.

Air ejectors will operate inefficiently if the ejector nozzles are coated or eroded. They should be inspected and cleaned or replaced regularly. The vacuum retaining valve should be checked for air tightness and also the ejector casing.

The various heat exchangers should be checked regularly for tube leakages and also the cleanliness of the heat-exchange surfaces. Turbo-feed pumps are started with the discharge valve closed in order to build up pressure rapidly and bring the hydraulic balance into operation. The turbine driving the pump will require warming through with the drains open before running up to speed and then closing the drains. The turbine overspeed trip should be checked regularly for correct operation and axial clearances should be measured, usually with a special gauge.

FUNCTION OF FEED PUMP IN A BOILER FEED SYSTEM/ STEAM TURBINES

THE FEED PUMP RAISES THE FEEDWATER

The feed pump raises the feedwater to a pressure high enough for it to enter the boiler. For auxiliary boilers, where small amounts of feedwater are pumped, a steam-driven reciprocating positive displacement pump may be used. Another type of feed pump often used on package boiler installations is known as an 'electro feeder'. This is a multi-stage centrifugal pump driven by a constant speed electric motor. The number of stages is determined by the feed quantity and discharge pressure.

Steam turbine-driven feed pumps are usual with high-pressure watertube boiler installations. A typical turbo-feed pump. The two-stage horizontal centrifugal pump is driven by an impulse turbine, the complete assembly being fitted into a common casing.

The turbine is supplied with steam directly from the boiler and exhausts into a back-pressure line which can be used for feed heating. The pump bearings are lubricated by filtered water which is tapped off from the first-stage impeller. The feed discharge pressure is maintained by a governor, and overspeed protection trips are also provided.

FUNCTION OF FEED EXTRACTION PUMP IN A FEED SYSTEM FOR BOILER AND STEAM TURBINE

HOW AN AUXILIARY FEED SYSTEM WORKS ?

The extraction pump is used to draw water from a condenser which is under vacuum. The pump also provides the pressure to deliver the feed water to the de-aerator or feed pump inlet. Extraction pumps are usually of the vertical shaft, two stage, centrifugal type.

These pumps require a specified minimum suction head to operate and, usually, some condensate level control system in the condenser. The first-stage impeller receives water which is almost boiling at the high vacuum conditions present in the suction pipe. The water is then discharged at a slight positive pressure to the second-stage impeller which provides the necessary system pressure at outlet.

Where the condenser sump level is allowed to vary or maintained almost dry, a self-regulating extraction pump must be used. This regulation takes the form of cavitation which occurs when the suction head falls to a very low value.

Cavitation is the formation and collapse of vapour bubbles which results in a fall in the pump discharge rate to zero. As the suction head improves the cavitation gradually ceases and the pump begins to discharge again. Cavitation is usually associated with damage but at the low-pressure conditions in the pump no damage occurs. Also, the impeller may be designed so that the bubble collapse occurs away from the impeller, i.e. super-cavitating.

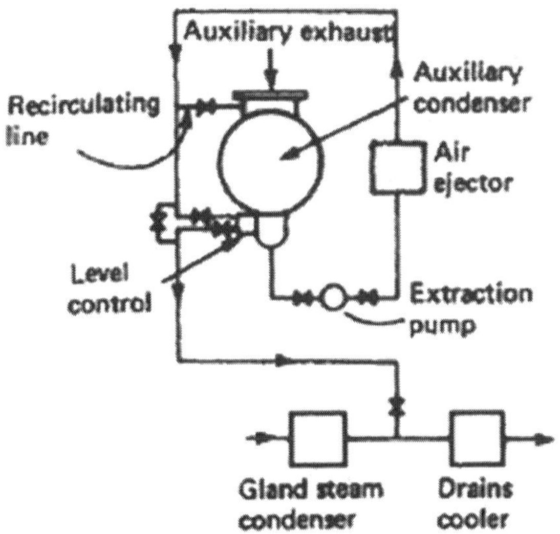

Fig. Auxiliary feed system

FUNCTION OF HIGH PRESSURE FEED HEATER IN BOILER FEED SYSTEM

HIGH-PRESSURE FEED HEATER

The feed pump raises the feedwater to a pressure high enough for it to enter the boiler.

The high-pressure feed heater is a heat exchanger of the shell and tube type which further heats the feedwater before entry to the boiler. Further heat may be added to the feedwater without its becoming steam since its pressure has now been raised by the feed pump.

The incoming feedwater circulates through U-tubes with the heating steam passing over the outside of the tubes. Diaphragm plates serve to support the tubes and direct the steam through the heater. A steam trap ensures that all the heating steam is condensed before it leaves the heater. Bled steam from the turbine will be used for heating.

FUNCTION OF HEAT EXCHANGERS IN A BOILER FEED SYSTEM/ STEAM TURBINES

HEAT EXCHANGERS

The gland steam condenser, drains cooler and low-pressure feed heater are all heat exchangers of the shell and tube type. Each is used in some particular way to recover heat from exhaust steam by heating the feedwater which is circulated through the units.

The gland steam condenser collects steam, vapour and air from the turbine gland steam system. These returns are cooled by the circulating feed water and the steam is condensed. The condensate is returned to the system via a loop seal or some form of steam trap and any air present Is discharged into the atmosphere. The feedwater passes through U-tubes within the shell of the unit.

The drains cooler receives the exhaust drains from various auxiliary services and condenses them: the condensate is returned to the feed system. The circulating feedwater passes through straight tubes arranged in tube plates in the drains cooler. Baffles or diaphragm plates are fitted to support the tubes and also direct the flow of the exhaust drains over the outside surface of the tubes

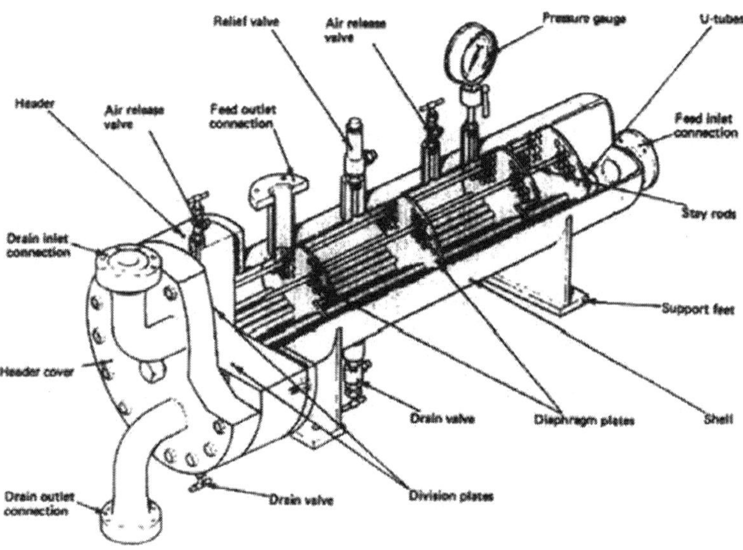

Fig. Feed system drains cooler

The low-pressure feed heater is supplied with steam usually bled from the low-pressure turbine casing. The circulating feed is heated to assist in the de-aeration process. The bleeding-off of steam from the turbine improves plant thermal efficiency as well as reducing turbine blade heights in the final rows because of the reduced mass of steam flowing. Either straight or U-tube construction may be used with single or multiple passes of feedwater.

HOW AN AIR EJECTOR WORKS IN A BOILER FEED SYSTEM

AIR EJECTOR

The air ejector draws out the air and vapours which are released from the condensing steam in the condenser. If the air were not removed from the system it could cause corrosion problems in the boiler. Also, air present in the condenser would affect the condensing process and cause a back pressure in the condenser.

The back pressure would increase the exhaust steam pressure and reduce the thermal efficiency of the plant.

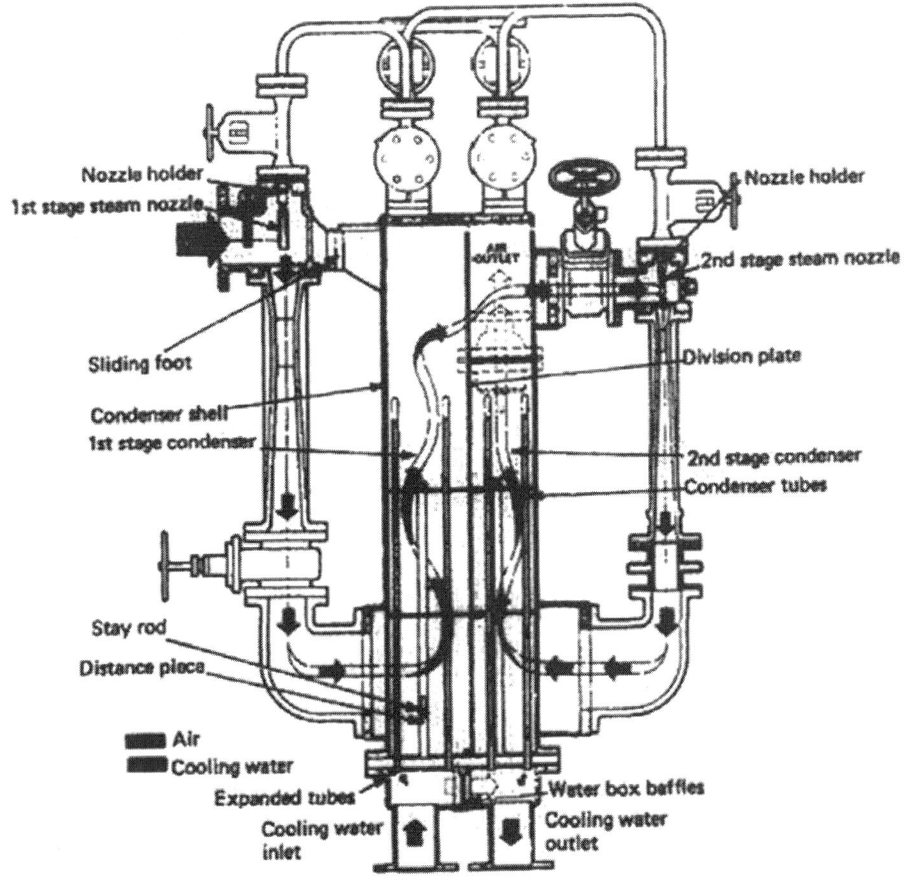

Fig. Feed system air ejector

In the first stage a steam-operated air ejector acts as a pump to draw in the air and vapours from the condenser. The mixture then passes into a condensing unit which is circulated by feedwater. The feedwater is heated and the steam and gases are mostly condensed.

The condensed vapours and steam are returned to the main condenser via a drain and the remaining air and gases pass to the second stage where the process is repeated. Any remaining air and gases are released to the atmosphere via a vacuum-retaining valve.

The feed water is circulated through U-tubes in each of the two stages. A pair of ejectors are fitted to each stage, although only one of each is required for satisfactory operation of the unit.

FUNCTION OF CONDENSER IN A FEED SYSTEM FOR BOILER AND STEAM TURBINE

THE CONDENSER AS A HEAT EXCHANGER

The condenser is a heat exchanger which removes the latent heat from exhaust steam so that it condenses and can be pumped back into the boiler. This condensing should be achieved with the minimum of under-cooling, i.e. reduction of condensate temperature below the steam temperature. A condenser is also arranged so that gases and vapours from the condensing steam are removed.

An auxiliary condenser. The circular cross-section shell is provided with end covers which are arranged for a two-pass flow of sea water. Sacrificial corrosion plates are provided in the water boxes.

The steam enters centrally at the top and divides into two paths passing through ports in the casing below the steam inlet hood. Sea water passing through the banks of tubes provides the cooling surface for condensing the steam.

The central diaphragm plate supports the tubes and a number of stay rods in turn support the diaphragm plate. The condensate is collected in a sump tank below the tube banks. An air suction is provided on the condenser shell for the withdrawal of gases and vapours released by the condensing steam.

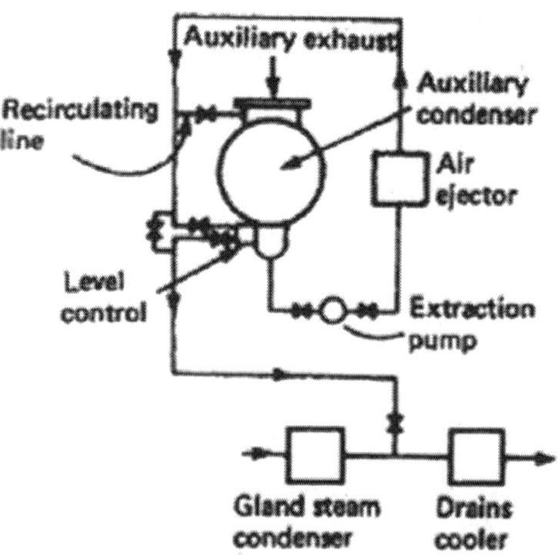

Fig. Auxiliary feed system

Main condensers associated with steam turbine propulsion machinery are of the regenerative type. In this arrangement some of the steam bypasses the tubes and enters the condensate sump as steam. The condensate is thus reheated to the same temperature as the steam, which increases the efficiency of the condenser. One design of regenerative condenser.

A central passage enables some of the steam to pass to the sump, where it condenses and heats the condensate. A baffle plate is arranged to direct the gases and vapours towards the air ejector. The many tubes are fitted between the tube plates at each end and tube support plates are arranged between. The tubes are circulated in two passes by sea water.

The feed system completes the cycle between boiler and turbine to enable the exhausted steam to return to the boiler as feedwater. The arrangements for steam recovery from auxiliaries and ship services may form separate open or closed feed sysems or be a part of the main feed system.

HOW DE-AERATOR WORKS IN A BOILER FEED SYSTEM

AIR AND VAPOUR REMOVAL PROCESS BY DE-AERATOR

The de-aerator completes the air and vapour removal process begun in the condenser. It also functions as a feed heater, but in this case operates by direct contact. The feedwater is heated almost to the point of boiling, which releases all the dissolved gases which can then be vented off.

One type of de-aerator. The incoming feedwater passes through a number of spray valves or nozzles: the water spray thus provides a large surface area for contact with the heating system. Most of the feedwater will then fall onto the upper surface of the de-aerating cone where it is further heated by the incoming steam. The feedwater then enters the central passage and leaves through a narrow opening which acts as an eductor or ejector to draw steam through with the feed.

The feedwater and condensed steam collect in the storage tank which forms the base of the de-aerator. The heating steam enters the de-aerator and circulates throughout, heating the feedwater and being condensed in its turn to combine with the feedwater. The released gases leave through a vent connection and pass to a vent condenser or devaporiser. Any water vapour will be condensed and returned. The devaporiser is circulated by the feedwater before it enters the de-aerator.

The de-aerator feedwater is very close to the steam temperature at the same pressure and will, if subjected to any pressure drop, 'flash-off into steam. This can result in 'gassing', i.e. vapour forming in the feed pump suction. To avoid this problem, the de-aerator is mounted high up in the machinery space to give a positive suction head to the feed pumps. Alternatively a booster or extraction pump may be fitted at the de-aerator outlet.

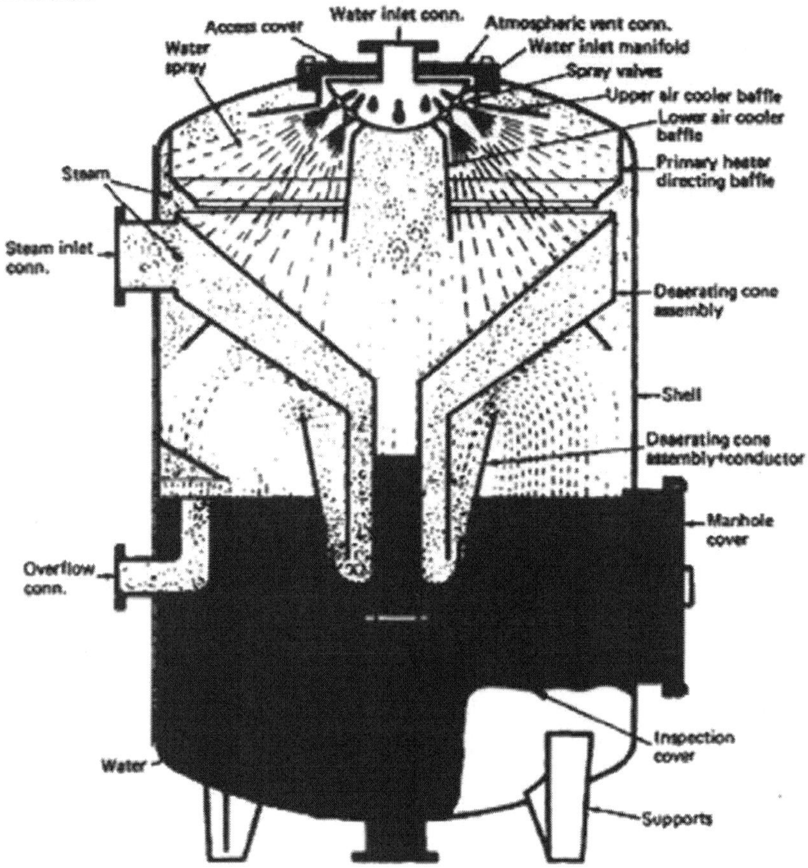

Fig. Feed system De-aerator

EVAPORATOR (MARINE)

This steam evaporator aboardHMS *Belfast* distilled up to six tons of fresh water per hour for the boiler and for drinking.

An evaporator, distiller or distilling apparatus is a piece of ship's equipment used to produce fresh drinking water from sea water bydistillation. As fresh water is bulky, may spoil in storage, and is an essential supply for any long voyage, the ability to produce more in mid-ocean is important for any ship.

EARLY EVAPORATORS ON SAILING VESSELS

Although distillers are often associated with steam ships, their use pre-dates this. Cook's Pacific exploration ship HMS *Resolution* of 1771 carried a distiller and Nelson'sHMS *Victory* of 1805 was fitted with distilling apparatus in her galley. Distilling apparatus was only fitted to larger warships and some exploratory ships at this time: a warship's large crew needed more water and they could ill afford the space to carry enough. Cargo ships, and their smaller crews, merely carried their supplies with them.

BOILER FEEDWATER

With the development of the marine steam engine, their boilers also required a continual supply of feedwater.

Early boilers used seawater directly, but this gave problems with the build-up of brine and scale. For efficiency, as well as conserving feedwater, marine engines have usually been condensing engines. By 1865, the use of an improved surface condenser permitted the use of fresh water feed, as the additional feedwater now required was only the small amount required to make up for losses, rather than the total passed through the boiler. Despite this, a large warship could still require up to 100 tons of fresh water makeup to the feedwater system, when under full power. Attention was also paid to de-aereating feedwater, to further reduce boiler corrosion.

The distillation system for boiler feedwater at this time was usually termed an evaporator, partly to distinguish it from a separate system or distiller used for drinking water. Separate systems were often used, especially in early systems, owing to the problem of contamination from oily lubricants in the feedwater system and because of the greatly different capacities required in larger ships. In time, the two functions became combined and the two terms were applied to the separate components of the system.

POTABLE WATER DISTILLERS

The first water supply by distillation of boiler steam appeared on early paddle steamers and used a simple iron box in the paddle boxes, cooled by water splash. A steam supply direct from the boiler, avoiding the engine and its lubricants, was led to them. With the development of steam heating jackets around the cylinders of engines such as the trunk engine, the exhaust from this source, again unlubricated, could be condensed.

EVAPORATORS

Combined supply

The first distilling plants that boiled a separate water supply from that of the main boiler, appeared around 1867. These are not directly heated by a flame, but have a primary steam circuit using main boiler steam through coils within a steam drum or *evaporator*. The distillate from this vessel then passes to an adjacent vessel, the *distilling condenser*. As these evaporators used a 'clean' seawater supply directly, rather than contaminated water from the boiler circuit, they could be used to supply both feedwater and drinking water. These *double distillers* appeared around 1884. For security against failure, ships except the smallest were fitted with two sets.

Vacuum evaporators

Evaporators consume a great deal of steam, and thus fuel, in relation to the quantity of fresh water produced. Their efficiency is improved by working them at a partial vacuum, supplied by the main engine condensers. On modern diesel-powered ships, this vacuum can instead be produced by an ejector, usually worked by the output from the brine pump. Working under vacuum also reduces the temperature required to boil seawater and thus permits evaporators to be used with lower-temperature waste heat from the diesel cooling system.

Scale

One of the greatest operational problems with an evaporator is the build-up of scale. Its design is tailored to reduce this, and to make its cleaning as effective as possible. The usual design, as developed by Weir and the Admiralty, is for a vertical cylindrical drum, heated by steam-carrying drowned coils in the lower portion. As they are entirely submerged, they avoid the most active region for the deposition of scale, around the waterline. Each coil consists of one or two spirals in a flat plane. Each coil is easily removed for cleaning, being fastened by individual pipe unions through the side of the evaporator. A large door is also provided, allowing the coils to be removed or replaced. Cleaning may be carried out mechanically, with a manual scaling hammer. This also has a risk of mechanical damage to the tubes, as the slightest pitting tends to act as a nucleus for scale or corrosion. It is also common practice to break light scaling free by thermal shock, by passing steam through the coils without cooling water present or by heating the coils, then introducing cold seawater.

Despite the obvious salinity of seawater, salt is not a problem for deposition until it reaches the saturation concentration. As this is around seven times that of seawater and evaporators are only operated to a concentration of two and a half times, this is not a problem in service. A greater problem for scaling is the deposition of calcium sulphate. The saturation point for this compound decreases with temperature above 60°C, so that beginning from around 90°C a hard and tenuous deposit is formed.

To further control scale formation, equipment may be provided to automatically inject a weak citric acid solution into the seawater feed. The ratio is 1:1350, by weight of seawater.

Compound evaporators

Operation of an evaporator represents a costly consumption of main boiler steam, thus fuel. Evaporators for a warship must also be adequate to supply the boilers at continuous full power when required, even though this is rarely required. Varying the vacuum under which the evaporator works, and thus the boiling point of the feedwater, may optimise production for either maximum output, or better efficiency. Greatest output is achieved when the evaporator operates at near atmospheric pressure and a high temperature (forsaturated steam this will be at a limit of 100°C), which may then have an efficiency of 0.87 kg of feedwater produced for each kg of steam supplied.

If condenser vacuum is increased to its maximum, evaporator temperature may be reduced to around 72°C. Efficiency increases until the mass of feedwater produced almost equals that of the supplied steam, although production is now restricted to 0.86 of the previous maximum.

Evaporators are generally installed as a set, where two evaporators are coupled to a single distiller. For reliability, large ships will then have a pair of these sets. It is possible to arrange these sets of evaporators in either parallel or in series, for either maximum or most efficient production. This arranges the two evaporators so that the first operates at atmospheric pressure and high temperature (the maximum output case), but then uses the resultant hot output from the first evaporator to drive a second, running at maximum vacuum and low temperature (the maximum efficiency case).

The overall output of feedwater may *exceed* the weight of steam first supplied, as up to 1.6 times as much. Capacity is however reduced, to 0.72 times the maximum.

Evaporator pumps

The unevaporated seawater in an evaporator gradually becomes a concentrated brine and, like the early steam boilers with seawater feed, this brine must be intermittently*blown down* every six to eight hours and dumped overboard. Early evaporators were simply mounted high-up and dumped their brine by gravity.

As the increasing complexity of surface condensers demanded better feedwater quality, a pump became part of the evaporator equipment. This pump had three combined functions as a seawater feed pump, a fresh water delivery pump and a brine extraction pump, each of progressively smaller capacity. The brine salinity was an important factor in evaporator efficiency: too dense encouraged scale formation, but too little represented a waste of heated seawater. The optimum operating salinity was thus fixed at three times that of seawater, and so the brine pump had to remove at least one third of the total feedwater supply rate. These pumps resembled the steam-powered reciprocating feedwater pumps already in service. They were usually produced by the well-known makers, such as G & J Weir. Vertical and horizontal pumps were used, although horizontal pumps were favoured as they encouraged the de-aeration of feedwater. Electrically powered rotary centrifugal pumps were later adopted, as more efficient and more reliable. There were initial concerns whether these would be capable of pumping brine against the vacuum of the evaporator and so there was also a transitional type where a worm gear-driven plunger pump for brine was driven from the rotary shaft.

FLASH DISTILLERS

A later form of marine evaporator is the flash distiller. Heated seawater is pumped into a vacuum chamber, where it 'flashes' into pure water vapour. This is then condensed for further use.

As the use of vacuum reduces the vapour pressure, the seawater need only be raised to a temperature of 77 °C (171 °F). Both evaporator and distiller are combined into a single chamber, although most plants use two joined chambers, worked in series. The first chamber is worked at 23.5 inHg (80 kPa) vacuum, the second at 26–27 inHg (88–91 kPa). Seawater is supplied to the distiller by a pump at around 20 psi. The cold seawater passes through a condenser coil in the upper part of each chamber before being heated by steam in an external feedwater heater. The heated seawater enters the lower part of the first chamber, then drains over a weir and passes to the second chamber, encouraged by the differential vacuum between them. The brine produced by a flash distiller is only slightly concentrated and is pumped overboard continuously.

Fresh water vapour rises through the chambers and is condensed by the seawater coils. Baffles and catchment trays capture this water in the upper part of the chamber. Vacuum itself is maintained by steam ejectors.

The advantage of the flash distiller over the compound evaporator is its greater operating efficiency, in terms of heat supplied. This is due to working under vacuum, thus low temperature, and also the regenerative use of the condenser coils to pre-heat the seawater feed.

A limitation of the flash distiller is its sensitivity to seawater inlet temperature, as this affects the efficiency of the condenser coils. In tropical waters, the distiller flowrate must be throttled to maintain effective condensation. As these systems are more modern, they are generally fitted with an electric salinometer and some degree of automatic control.

VAPOUR-COMPRESSION DISTILLERS

One of two vapour-compression distillers in the engine room of WW2 submarine USS Pampanito (SS-383)

Diesel-powered motorships do not use steam boilers as part of their main propulsion system and so may not have steam supplies available to drive evaporators. Some do, as they use auxiliary boilers for non-propulsion tasks such as this. Such boilers may even be heat-recovery boilersthat are heated by the engine exhaust.

Where no adequate steam supply is available, a vapour-compression distiller is used instead. This is driven mechanically, either electrically or by its own diesel engine.

Seawater is pumped into an evaporator, where it is boiled by a heating coil. Vapour produced is then compressed, raising its temperature. This heated vapour is used to heat the evaporator coils. Condensate from the coil outlet provides the fresh water supply. To start the cycle, an electric pre-heater is used to heat the first water supply. The main energy input to the plant is in mechanically driving the compressor, not as heat energy.

Both the fresh water production and the waste brine from the evaporator are led through an output cooler. This acts as a heat exchanger with the inlet seawater, pre-heating it to improve efficiency. The plant may operate at either a low pressure or slight vacuum, according to design. As the evaporator works at pressure, not under vacuum, boiling may be violent. To avoid the risk of priming and a carry over of saltwater into the vapour, the evaporator is divided by a bubble cap separator.

Submarines

Vapour-compression distillers were installed on US submarines shortly before World War 2. Early attempts had been made with evaporators running from diesel engine exhaust heat, but these could only be used when the submarine was running at speed on the surface. A further difficulty with submarines was the need to produce high-quality water for topping up their large storage batteries. Typical consumption on a war patrol was around 500 US gallons per day for hotel services, drinking, cooking, washing etc. and also for topping up the diesel engine cooling system. A further 500 gallons per week was required for the batteries. The standard Badger model X-1 for diesel submarines could produce 1,000 gallons per day. Tank capacity of 5,600 gallons (1,200 of which was battery water) was provided, around 10 days reserve. With the appearance of nuclear submarines and their plentiful electricity supply, even larger plant could be installed. The X-1 plant was designed so that it could be operated when snorkelling, or even when completely submerged. As the ambient pressure increased when submerged, and thus the boiling point, additional heat was required in these submarine distillers and so they were designed to run with electric heat continuously.

Bibliography

Akber Ayub: *Marine Diesel Engines*, Ane Books, Delhi, 2015.

Alok Tripathi: *Maritime Heritage of Indian Ocean*, Sharada Publishing House, Delhi, 2013.

Ananta Narayan Mishra: *Glimpses of Maritime History of Ancient Odisha*, Kunal Books, Delhi, 2010.

Bell: *Diesel Engineering : Electricity and Electronics*, Cengage, 2009.

Bennett: *Diesel Engineering : Brakes, Steering and Suspension Systems*, Cengage, 2009.

Bennett: *Diesel Engineering : Electronic Diesel Engine Diagnosis*, Cengage, 2009.

Dixon: *Diesel Engineering : Heating, Ventilation, Air Conditioning and Refrigeration*, Cengage, 2009.

Gadepalli Ravi Kiran Sastry: *Bio-Diesel : Biodegradable Alternative Fuel for Diesel Engines*, Readworthy, 2008.

Gert Jan Bom, I H Rehman, David van Raalten, Rajeshwar Mishra and Frank van Steenbergen: *Technology Innovation and Promotion in Practice: Pumps, Channels, and Wells. Reducing Fuel Consumption, Emissions, and Costs*, TERI, 2002.

Gurpreet S Khurana: *Maritime Forces in Pursuit of National Security : Policy Imperatives for India*, Shipra Books, Delhi, 2008.

K. R. Singh: *Coastal Security : Maritime Dimensions of Indias Homeland*, Vij Books, Delhi, 2012.

K. Sridharan: *India's Maritime Heritage*, CBS Publication, Delhi, 2003.

K.M. Srinivasan: *Centrifugal Pumps*, New Age International, Delhi, 2016.

K.S. Pavithran: *Foreign Policy and Maritime Security of India*, New Century Publications, Delhi, 2013.

K.S.Mathew: *Imperial Rome, Indian Ocean Regions and Muziris: New Perspectives on Maritime Trade*, Manohar Publication, Delhi, 2015.

L.N. Swamy: *Maritime Contacts of Ancient India : With Special Reference to West Coast*, Harman Publication, Delhi, 2000.

Manoj Gupta: *Maritime Affairs : Integrated Management for India*, Manas Publication, Delhi, 2005.

MICHAEL: *Water Wells & Pumps, Second Edition*, Tata McGraw-Hill, 2008.

Pius Malekandathil: *Maritime India Trade, Religion and Polity in the Indian Ocean*, Primus Books, Delhi, 2010.

Pradeep Kaushiva and Kamlesh K. Agnihotri: *Indian and American Perspectives on Technological Developments in the Maritime Domain and Their Strategic Implications in the Indian Ocean Region*, KW Publishers, Delhi, 2013.

Ravi Vohra, P.K. Ghosh and D. Chakraborty: *Contemporary Transnational Challenges: International Maritime Connectivities (Indian Ocean Naval Symposium)*, KW Publication, Delhi, 2008.

Sila Tripati, A.S. Gaur and Sundaresh: *Maritime Archaeology and Shipwrecks off Goa*, Kaveri Books, Delhi, 2014.

Sila Tripati: *Maritime Archaeology for Beginners*, Kaveri Books, Delhi, 2009.

Sila Tripati: *Maritime Contacts of the Past: Deciphering Connections Amongst Communities*, Kaveri Books, Delhi, 2015.

Sugandha: *Evolution of Maritime Strategy and National Security of India*, Decent Books, Delhi, 2008.

V.S. Lobanoff and R. Ross: *Centrifugal Pumps : Designs and Application*, Jaico Publishing House, Delhi, 2005.

V.S. Lobanoff and R. Ross: *Centrifugal Pumps: Designs and Application*, Jaico Publishing House, Delhi, 2003.

Vijay Sakhuja: *Asian Maritime Power in the 21st Century : Strategic Transactions: China India and Southeast Asia*, Pentagon Press, New York, 2011.

Yogendra Kumar: *Diplomatic Dimension of Maritime Challenges for India in the 21st Century*, Pentagon Press, New York, 2015.

Index

I

L

M

O

P

R

S

T

V

W